THE BRIDGES OF NEW YORK

Sharon Reier

DOVER PUBLICATIONS, INC.
Mineola. New York

3rd Avenue Bridge c. 1864 COURTESY OF THE NEW YORK HISTORICAL SOCIETY

Front Cover—"59th Street Bridge"— WATERCOLOR BY HOWARD KALISH

Copyright

Published in Canada by General Publishing Company, Ltd., 30 Lesmill Road, Don Mills, Toronto, Ontario.

Bibliographical Note

This Dover edition, first published in 2000, is an unabridged reprint of the first edition of *The Bridges of New York,* published by Quadrant Press, New York, in 1977.

Library of Congress Cataloging-in-Publication Data

Reier, Sharon.
The bridges of New York / by Sharon Reier.
p. cm.
Originally published: New York : Quadrant Press, 1977.
Includes bibliographical references.
ISBN 0-486-41230-X (pbk.)
1. Bridges—New York (State)—New York. I. Title.

TG25. N5 R45 2000
624'.2'09747—dc21

00-031780

Manufactured in the United States of America
Dover Publications, Inc., 31 East 2nd Street, Mineola, N.Y. 11501

CONTENTS

Queensboro Bridge COURTESY OF THE NEW YORK TRANSPORTATION ADMINISTRATION

PREFACE

I wrote this book for several reasons. New York is my city. I was born here and have lived here most of my life. My family got its first car when I was three and we soon moved over the Triborough Bridge into the farthest reaches of Queens, where we became a typical suburban, media-saturated, slightly cultured, lawn-mowing family.

However, we frequently traveled back to the inner city to visit our vast network of grandmothers, grandfathers, uncles, aunts and cousins. During our trips we inevitably crossed a bridge, and that quickly became my favorite part of the journey—picking a bridge and then swooping over it.

When I grew older I wondered why some bridges were built one way and some another, and who built them and whether they were really safe and would they fall down? I began to collect information and read and I became fascinated by these awesome structures as I found out that New York's were the world's greatest.

While I was writing this book I was frequently warned that a treatment of all of New York's bridges already existed. I tried to track down this work and I failed. I think the confusion arose from books about modern bridgebuilding. Many of New York's spans were breakthroughs in engineering history, and therefore are included in such works.

However, I did not intend to write an engineering treatise for laymen. I merely wanted to explain why New York's bridges were designed the way they were. I found it impossible to separate the story of the bridges from the political and economic life of the city, or from modern developments in engineering. So, I have included nearly 300 years of New York history in this book, as well as a century of modern engineering history.

The organization of the book was somewhat difficult. Had I arranged the bridges in a chronological sequence, I would have had to jump from one river to another, tangling up the engineering problems as I proceeded. So it seemed simpler, and more in keeping with engineering practice to group the bridges by geography and economics. These are two of the most important problems investigated by an engineer about to build a bridge.

In writing the book I have received encouragement and help from a great many people and organizations, friends and relatives. I would like to thank the Port Authority of New York and New Jersey and New York Department of Transportation for their time and the facilities they put at my disposal. I also thank The Bronx County Historical Society, which gave me a free hand with its collection, as did Loren McMillen at the Staten Island Historical Society. As a former engineer and architect, Mr. McMillen was kind enough to familiarize me with several engineering principles.

Another debt is owed to August Rochlitz, a former Department of Transportation employe who permitted me to use his photos and materials and directed me to the facility at the Williamsburg Bridge. Also, I thank Harold Samuelson, a very fine civil engineer who enabled me to inspect the Manhattan Bridge and who read over parts of my manuscript for technical errors.

Marie Spina, assistant at the Brooklyn collection of the Brooklyn Public Library is responsible for finding many of the older photographs in this book, which were reproduced by Diana Weingar. Another library which gave excellent service and personal help was the New York City Municipal Research Library. As a private research source, the Broadway Limited, owned by Howard and Suzanne Samelson, was generous in providing material on railroad bridges.

More thanks go to Tom Flagg, my editor, and very special thanks to Gale Olson, who cleaned up some of my muddier prose and encouraged me when necessary.

I am also in great debt to my uncle, Sam Fagin, a fine engineer who was supportive and helpful and had faith in my project, to my parents who helped when they could, and to Reginald Dunlap, who corrected, criticized, typed and made me do things better.

Bayonne Bridge, the world's longest steel arch, is shown at dusk with its wreath of night lights.

INTRODUCTION

Over the waterways of New York City there are more than 75 bridges. Although some cities claim more bridges, nowhere are there so many vast steel monuments, built by the industrial era's foremost bridge builders.

For over a century New York has been the dominant city in the Western Hemisphere, acting as a magnet for creative people of all types. The same deep harbor and wide rivers which made New York a commercial success inspired the world's most talented and visionary engineers to establish a new scale in bridgebuilding here. The Seine and Thames, after all, are mere streams compared with New York's East and Hudson Rivers. New York's large spans would receive unparalleled recognition. In turn these bridges helped to knit the city together.

In 1964, when the Verrazano-Narrows Bridge was completed, New York regained its title for the world's longest suspension bridge. The Brooklyn, Williamsburg and George Washington bridges were earlier holders of this record. On the outskirts of the city stands the Bayonne Bridge, the world's longest steel arch. It is 25 inches longer than its nearest rival in Sydney, Australia. The Hell Gate Railroad Bridge also holds a record. It is the world's strongest steel arch. And the George Washington Bridge is still the world's heaviest suspension bridge.

A bridge, large or small, is composed of a harmonious combination of curves and angles, bolts and beams, wires and pins, steel and masonry. It must endure its own weight as well of that of passing traffic. It must be flexible, allowing for traffic and weather conditions. In the case of New York's large spans, there were no computers to determine the near-mystery of stresses and strains, size of pins or type of steel needed. The structures are the result of months and sometimes years of calculations by geniuses and near-geniuses whose originality was matched by their meticulousness and determination.

Bridges are a highpoint of any civilization and attract the finest minds. Although the names Gustav Lindenthal, John Roebling, Othmar Ammann and David Steinman, are not commonly known, we may equate their skills with Leonardo da Vinci's and Michelangelo's. Leonardo offered to build a masonry arch bridge over the Bosporus, with an unprecedented span of 905 feet. Michelangelo was a consultant on the Trinita Bridge in Florence. (This one was a casualty of World War II.) Like their Renaissance counterparts, New York's bridge builders were not limited in their talents. David Steinman wrote books and poetry. Gustav Lindenthal had great insight as a sociologist and economist. John Roebling was a philosopher, inventor, musician and linguist.

"All the bridge designers I've ever met were egomaniacs," commented a Canadian professor of civil engineering. "They have to be. You see, you can never really tell whether a bridge will stand until the last bolt is put in place, the roadway completed. Until then, it's just a matter of conviction."

When one reads the writings of New York's bridge builders, one is impressed by their unshakable confidence and the range of their minds. One cannot underplay the significance of the public's faith in the bridge builder. A novelist or composer may present a poor work to the public at one time and make a spectacular comeback later on. This is hardly true of a bridge builder. If his work fails, no city or corporation will invest in his next project.

Like most structures, bridges reflect the spirit of their age. The extraordinary size and grandeur of New York's bridges are consistent with the effusive philosophy of the past age: that bigger is better and that technology can be used to serve and grace humanity.

Many of the artists, writers and dramatists who flocked to New York have been impressed by the city's spans and have portrayed them in their works. But perhaps the bridges themselves are works of art. As Marcel Duchamp said: "The only contribution America has made to art are her plumbing and her bridges."

To really appreciate the city's bridges it is best to avoid auto routes. Automobile routes rarely show off their dramatic beauty. One extremely pleasant way to view them is from the waters below. This is true whether one is piloting one's own craft or riding on one of the excursion boats which cruise around Manhattan. However, sailors who wish to make the trip through the Harlem River in a high-masted boat should stand warned that seven of the river's thirteen bridges must be opened to accommodate the spar. One skillful yachtsman who found himself unable to buck the East River tides was forced into the Harlem River in order to complete a passage to Lower New York Bay. He was relieved, he said, that the city was not going to charge him for opening all the bridges, but he admitted that he also felt embarrassed by the time the fourth bridge was opened and a police escort was accompanying him up the strait to handle traffic jams.

Recently, a new city regulation makes it necessary to warn the city in advance of such openings. The Transportation Department, forced to cut back on manpower, now uses fewer bridgetenders and must arrange for operators to be on duty.

Of course, the most intimate way to explore and befriend a bridge is on foot. All the bridges in New York except four—the Verrazano-Narrows Bridge, the Throgs Neck and Whitestone Bridges, and the Alexander Hamilton Bridge across the Harlem River—were designed with footpaths, and those still in existence make for a breezy summer walk. They also afford some of the best views of the New York skyline.

If there is an end to the building of great bridges, at least in the New York Metropolitan area, it may be due to the eagerness of big cities to rid themselves of pollution producing autos. Over a decade ago, Robert Moses, the moving force behind many of New York's bridges, proposed a bridge to cross Long Island Sound. This idea has elicited little support among Long Island residents. Perhaps this is because in much recent bridge and highway design the needs of the people seem to be ignored and the bridges become expensive highway extensions serving cars and trucks, not people. It is ironic that less than a century ago Gustav Lindenthal designed recreational areas—cancelled by City Hall—around the towers of the Manhattan Bridge and that there was a constant promenade of people across the Brooklyn and Williamsburg Bridges.

The city has never had a bridge calamity, and, with proper maintenance, the spans may last for centuries, authorities say. As the bridges and aqueducts of ancient Rome served to teach our bridge-builders, so New York's bridges may inspire the builders of some future civilization.

Crossing over the ice to Manhattan before the Brooklyn Bridge was built. COURTESY OF THE BROOKLYN MUSEUM

THE EAST RIVER
NO MORE COWS IN BROOKLYN

The East River has long been known to navigators as one of the more hazardous passages on the continent. It is not a river, after all, but a broad tidal channel with a restless current that reverses with the rise and fall of the tide. For over 200 years it served as a boundary between two rival cities: Brooklyn and New York.

In fact, by the end of the Civil War, Brooklyn was America's third largest city. New York ranked first. Considerable commerce existed between the two cities but it was balanced by a serious antagonism. Brooklyn had a provincial image and called itself "City of Homes and Churches." New York was already cast as a center for tenements, crime and corrupt politicians. Many Brooklynites believed their city had all the advantages and saw no reason why it could not be bigger, richer and more successful than New York. Brooklyn did have the longer shoreline and its boatyards were busy.

However, the two cities had engaged in commerce from their inception. A Dutch farmer, Cornelius Dircksen, started a fairly regular passenger service as early as 1659. He charged a fee of three wampum to row a traveler across the river in his skiff. To gain transport, a passenger would ring a bell at the ferry slip to bring Dircksen from his fields. Weather and currents made the trip possible about once every three days, and eventually Dircksen sold out his business to a professional proprietor, who made more regular crossings, and a community sprang up around the ferry slip.

Sailboats called pirogues were running from Catherine Street, Manhattan, to Brooklyn by the time the Revolution started. These carried horse-drawn vehicles and livestock as well as passengers. Still, they could only run when the wind was right. Barges which limited their freight to foot-passengers ran more regularly, but they were considered unsafe.

In 1807 Brooklynites were finally given a secure means of crossing the East River. Robert Fulton's *Clermont* made its maiden voyage up the Hudson and a regular steamboat service began between Brooklyn and New York, as well as between New York and New Jersey. For three years Fulton and his mentor Robert Livingston held exclusive rights to run this service. A ride across the East River on the *Nassau* took five to 12 minutes, according to the tide. The ferries were odd-looking roundish tubs, but they were efficient and fairly safe. New residents flocked to Brooklyn. They found pleasure in working in New York and enjoying its advantages and then returning to comfortable safe communities. Fulton's boats were even hired out for pleasure parties.

By the time the first East River bridge was built, the ferries were carrying more than 120,000 passengers over the channel each day. When construction started on the Williamsburg Bridge, the ferries still were carrying that number of passengers and the Brooklyn Bridge was carrying an additional 150,000 people a day.

The main ferry line was the Union Ferry Company, which ran five lines—the Fulton Street, to Fulton; the Catherine Street, to Main; the Wall Street to Joralemon Street; the Whitehall Street to Atlantic Avenue, and the Whitehall Street to Hamilton Avenue.

It is said that the building of the Brooklyn Bridge, the first over the East River, may have been the decisive factor in the union of the two cities. It certainly sped up Brooklyn's growth.

Sentiment for a bridge had been expressed by prominent Brooklynites since before 1800. At that time General Jeremiah Johnson, who later became Brooklyn's mayor, had written in his journals, "It has been suggested that a bridge should be constructed across the East River to New York. This idea has been treated as chimerical, from the magnitude of the design; but whosoever takes it into their serious consideration will find more weight in the practicability of the scheme than at first sight he imagines."

In 1811, Thomas Pope, an architect and landscape gardener, published a book called *A Treatise on Bridge Architecture, In Which the Superior Advantages of the Flying Pendant Lever Bridge are Fully Proved.* In it, he proposed to build a wooden cantilever arch bridge across either the Hudson or East River. The book, which deals with the practical aspects of building such an enormous bridge, ends, oddly enough, in a verse essay of 105 heroic couplets celebrating the proposed bridge. The bridge was to have been a very flat arch, consisting of twin cantilevers joined at the center and stiffened with diagonal bracing. Pope claimed that such a structure could reach 3,000 feet across the Hudson. But, as a start, he built a 94 foot model to exhibit in New York. He claimed it could be reproduced on a larger scale to cross the East River. The bridge became popularly known as Pope's "rainbow bridge."

Another proposal was advanced in 1836 by General Joseph Swift, a military engineer. Swift had refurbished for peacetime use the old line of entrenchments which Washington's Revolutionary Army had used, located on dikes between Wallabout, Brooklyn, and Gowanus. Swift's idea was to construct a dike across the East River upon which a boulevard would be laid out.

The hope for a bridge prompted Brooklyn to dub one street near the waterfront "Bridge Street." This street, near the terminals of the Manhattan and Brooklyn Bridges, was named long before either bridge was planned.

None of the early proposals suggested the technology which the crossing actually required. The wooden construction of the Pope bridge, even if technically feasible, would have been suitable only for pedestrians. Besides having the same drawback, Swift's plan would have impaired the navigability of the busy waterway.

In fact, a New York and Brooklyn Bridge Company Report mentions that by the late 1860s as many as 100 ships passed the bridge site within an hour.

It was not until after the Civil War that technology advanced enough to allow for the Brooklyn Bridge, which had to provide for the daily crossing of the city's masses of people, either by carriage or by train.

Today there are six bridges crossing the East River. Four of them, the Brooklyn Bridge, the Williamsburg Bridge, the Queensboro Bridge and the Manhattan Bridge—in order of construction—were built before World War I and were considered great feats of engineering. They are all under the jurisdiction of the New York Department of Highways. There is no toll charge on any of the four.

Each of these bridges has its own engineer to inspect daily for weaknesses and other problems. They were all built with pathways for pedestrians and this is probably the best way to appreciate them. In recent years Mayor John Lindsay furnished each bridge with a string of decorative night lights. The lights not only illuminate the harbor but add a luster to these triumphant structures.

The Brooklyn Bridge and the Williamsburg Bridge originally had tolls for carriage crossings and the Brooklyn Bridge even charged a penny to pedestrians. These tolls were removed in 1911 under Mayor William Gaynor's administration. Gaynor took office just as the Queensboro and Manhattan Bridges were finished. Most of the construction on these two bridges was accomplished under the administration of Tammany Mayor George McClellan, Jr.

A recent solution to the mass transit problems of New York, proposed in 1968, called for collecting tolls once again on these four East River bridges. It was defeated as the cost of setting up and manning toll booth plazas seemed prohibitive. The fifth span across the East River, the Triborough, is a toll bridge. Another bridge, The Hell Gate, was built by railroad interests and only accommodates trains.

PHOTO BY PETER ROSE

BROOKLYN BRIDGE

THE GREAT EXPERIMENT

The Brooklyn Bridge was the world's first steel suspension bridge. Its span—1595.5 feet between towers—made it the world's longest. But it would not have such universal appeal if it were not for the impressive appearance of its solid granite towers with their Gothic arches and the fine steel webwork which suspends the roadway. One controversy surrounding the sedate, gleaming structure is whether it is better viewed from the shores of Manhattan or from its Brooklyn side. Workers at New York's Municipal Building, which looks out upon the bridge, contend that from Manhattan one can contemplate the perfect form of the serene harp strung structure as it soars into the amorphous maze of Brooklyn neighborhoods. Brooklynites seem to prefer the more vibrant quality of the bridge with the skyscrapers of lower Manhattan etched against the sky.

John Augustus Roebling, engineer, visionary, spiritualist, metaphysician, inventor and businessman designed the bridge while living on the Brooklyn side at 37 Hicks Street in Brooklyn Heights. His son, Colonel Washington Roebling, who took over as chief engineer after his father's death, also lived in Brooklyn at 110 Columbia Heights. From his window Washington Roebling could watch the masterpiece under construction. Living in these same quarters some time later poet Hart Crane claimed his imagination was fired by the sight of the bridge and wrote his famous poem to it.

The late David Steinman, who became one of America's most eminent bridgebuilders, attributed his ambition to the impact the Brooklyn Bridge had upon him. As a child Steinman lived in a tenement on the Lower East Side of Manhattan where he could see the impressive span from his bedroom window. He grew up to design the Florianapolis Bridge in Brazil, the Mount Hope Bridge in Rhode Island, the Grand Mere in Quebec, the Henry Hudson steel arch bridge across the Spuyten Duyvil Creek in New York, and was later chosen to modernize the bridge he most admired. In 1948 Steinman designed braces to be added over the traffic lanes of the Brooklyn Bridge which gave it greater strength.

As a tribute to the prophetic force the Roeblings had on the subsequent construction of bridges and the debt he personally felt, Steinman wrote *The Builders of the Bridge,* a painstakingly researched biographical work on John and Washington Roebling.

John Roebling, a German immigrant born in 1806, had seen his first suspension bridge during his student days at the Royal Polytechnic Institute in Berlin. It was a small structure which hung from iron chains over the Regnitz stream in the town of Bamburg, Bavaria. It was no match for the heroic spans Roebling envisioned himself designing.

He was well trained to fulfill his fantasies. Although he had come from the family of a poor, small town tobacconist, Roebling's mother was ambitious. Recognizing his intelligence and diligence, she secured for him an excellent education. His university studies included mathematics, engineering, architecture, languages, history and philosophy. During his university career he became a protege of the brilliant, charismatic philosophy professor Georg Hegel. Roebling was deeply influenced by the idealistic philosopher, and it may have partly been Hegel's enthusiasm for the American ideals of freedom and democracy that brought the young engineer to this country. Roebling's 2000 page treatise on his concept of the universe, which covers metaphysics, spiritualism, aesthetics and politics as well as mathematics and engineering, indicates the extent to which he absorbed the philosopher's teachings.

Alan Trachtenberg, an academician who explored the philosophy behind the creation of the Brooklyn Bridge in his book *Brooklyn Bridge, Fact and Symbol,* explains that Roebling considered physical matter as an expression of spirit, thinking of his suspension bridges as monuments to the spirit of progress and the harmony of nature. According to Trachtenberg, Roebling saw his bridge as a "world historical object" in Hegel's frame of reference.

It would probably come as no surprise to the utopian apostle of steel and engineering that his bridge has excited so many and has become a monument. After all, when presenting the idea in the Report to the New York Bridge Company in 1867, Roebling wrote:

> "The contemplated work, when constructed in accordance with my design, will not only be the greatest bridge in existence, but it will be the greatest engineering work of this continent, and of the age. Its most conspicuous features, the towers, will serve as landmarks to the adjoining cities, and they will be entitled to be ranked as national monuments."

These grandiose claims were not the idle speculations of a dreamer. They are the words of an already successful man with a practical but expensive idea to promote. By 1867 Roebling had already constructed two impressive suspension bridges: The Niagara Suspension Bridge, the first successful railroad suspension bridge in the world, and the Covington Bridge across the Ohio River from Cincinnati to Covington, Kentucky, the longest suspension bridge up to that time. In addition, he had earlier built several suspension aqueducts over the Allegheny River. Despite his successes, Roebling had found each new project met with skepticism, ignorance, politicking, and worst of all, panic. He had learned that it was not enough to be merely inventive and competent. He had to write pamphlets, indoctrinate the public, overcome the opposition, sell himself.

His previous projects paled in comparison with his vision of the great East River span. Originally he had conceived the bridge in 1852 while his first bridge was just getting under way at Niagara. It took over fifteen years for him to get approval for the East River Bridge, the achievement of which was to cause his death and ruin his son's health.

By the time John Roebling emigrated to America he had earned a degree in civil engineering and had worked for three years building roads and bridges for the Prussian government. He found the work uncreative. He felt stifled by Prussian authorities, and it is believed the government considered him a subversive for advocating mass emigration. It was 1831 and there had been a number of minor revolutions throughout Europe. When he left, Roebling had to sneak out of the country.

He had virtually abandoned the idea of becoming a bridge-builder, and leaving his engineering ambitions behind, planned to establish a farming colony in the new world with his brother and others

Right: Dark suited gentlemen stroll and rest on the promenade of the Brooklyn Bridge in the 1880s. Even then it provided a respite from the busy city. Note that at the time, the walkway's wooden planking was laid longitudinally. There was also a toll booth at the tower.

Below: Today the promenade is still a delightful place to walk and observe the imposing profile of Lower Manhattan. Its boardwalk is now laid horizontally and no tolls are charged.

COURTESY OF THE BROOKLYN PUBLIC LIBRARY BROOKLYN COLLECTION

PHOTO BY PETER ROSE

emigrating from his native town of Thuringen. The colony, called Germania at first and then Saxonburg, was near Pittsburgh and served as Roebling's home for six years.

Although Roebling devoted much of his energy to agriculture and building homes for more newly arrived immigrants, he spent time playing flute and piano, writing, and studying engineering. He was known as a man who was extremely prudent with his time and energies. One story David Steinman tells is that on an occasion during the Civil War, General Fremont sent for Roebling and kept him waiting in the anteroom. Finally, growing impatient, the engineer sent in a card with these words: "Sir, I am happy to do any work you want. But waiting in idleness is a luxury I never permit myself." It was said that if a man was five minutes late to an appointment, Roebling refused to see him.

By 1837 Roebling had become a United States citizen, had married and become the father of three sons, one of whom was to complete Roebling's greatest work. John Roebling decided he didn't want to devote his life to farm work. Living in Western Pennsylvania had shown Roebling the necessity for improved transportation of all types on the vast new continent, and he took a job as assistant engineer on the Beaver River Canal, then worked on the Allegheny River, constructing feeders for the Pennsylvania Canal. He then turned to surveying work, which led to the Pennsylvania Railroad's construction of a route between Harrisburg and Pittsburgh.

It was during this period of his life that Roebling developed the wire rope which was to make the construction of his suspension bridges possible, and which is still used today.

While working on the Harrisburg-Pittsburgh route, Roebling observed that the hemp ropes which hauled canal boats up and down mountains frequently broke, sometimes causing tragic accidents. Roebling recalled a paper he had read while a student which dealt with the making of wire rope. He was unable to remember the details, but set out to reproduce the idea through his own ingenuity. He set up a "rope walk" in Saxonburg, purchased iron wire and taught his neighbors to twist the wire into cables. This was the first wire cable to be manufactured in America. It immediately won favor in the portage of canal barges and Roebling began to develop a reputation.

Also at that time in his life—around 1841—Roebling learned that a certain flamboyant engineer named Charles Ellet, a young American who had studied bridgebuilding in Europe, had proposed the construction of a suspension bridge across the Schuylkill River in Philadelphia. Ellet had returned from Europe wild-eyed with the conviction that America must have suspension bridges. He proposed to Congress that he be commissioned to build a suspension bridge of over 1000 feet spanning the Potomac. The idea was considered outrageous. So it was on the Philadelphia bridge that Ellet had to prove the feasibility of his scheme Roebling wrote a letter to Ellet, telling of his study of bridgebuilding, his particular interest in suspension bridges, expressing enthusiasm and the desire to aid Ellet on the project. "Let but a single bridge of the kind be put up in Philadelphia exhibiting all the beautiful forms of the system to the best advantage, and it needs no prophecy to foretell the effect, which the novel and useful features will produce upon the intelligent minds of the Americans. You will certainly occupy a very enviable position, in being the

COURTESY OF THE BROOKLYN PUBLIC LIBRARY BROOKLYN COLLECTION

Engineer and inventor John Roebling had strong mystical and philosophical leanings. He died before building his bridge.

first engineer who, aided by nothing but the resources of his own mind and a close investigation, succeeded in introducing a new mode of construction, which here will find more useful application than in any other country," Roebling wrote. He received a polite reply but was never further contacted.

Ellet's Schuylkill Bridge was 358 feet long and was supported by five wire cables on a side. It was the weaker type of cables Ellet used which account for his subordinate place in the history of bridgebuilding. In fact, five years after Ellet built the world's first long-span wire cable suspension bridge —the 1010 foot Wheeling Bridge across the Ohio River—Roebling was hired to repair and strengthen it, as it had been badly damaged in a storm.

Ellet's cables were composed of parellel strands of wire connected by iron bars, from which the suspenders were hung. Roebling's cables were parallel strands compressed together hexagonally to form compact cylinders which were wrapped with light wire for protection against weather. The suspenders were hung from iron clamps that encircled the cables.

Roebling had his first opportunity to prove the usefulness of his cables not on a bridge, but on suspended aqueducts over the Allegheny River. The traditional type of aqueduct on which canal boats went across the river was built on piers that were easily damaged by fast flowing currents or ice in freezing weather. Roebling proposed he be permitted to try his new concept of supended aqueducts. The untried idea was at first considered preposterous, but as he gave a very low estimate of cost, he was permitted to go ahead with the project. It was so successful he was asked to build several more. One of the aqueducts, at Lackawaxen, Pennsylvania, is still standing, and has been converted to use as an automobile crossing. It is now the only major privately owned suspension bridge in the country.

From this success Roebling was convinced that his cables would have value and began to take steps to realize a destiny he now saw clearly before him. He established the Roebling works in Trenton, New Jersey, still the world's largest manufacturer of wire cable. He published a monograph in the *American Railroad Journal* on the superiority of wire cable over iron chain suspension bridges, discussing the weakness of earlier suspension bridges and their vulnerability to destructive cumulative vibration caused by winds.

His name was so respected in engineering circles by this time that when a railroad suspension bridge over the Niagara River was planned, Roebling's opinion was sought, along with Charles Ellet's. Roebling seriously set about doing calculations and wrote a sincere, convincing testimony to the feasibility of such a bridge. But the sensation-seeking Charles Ellet was chosen to do the building. Ellet had hired promoters to get his name into the newspapers in connection with the new bridge. He offered to subscribe to a large portion of shares in the bridge company. He realized, as did Roebling, that the man who could successfully complete such a structure would gain the admiration of the world.

To show the imagination Ellet had, which seems to be a necessary attribute to bridge builders, Ellet offered a $10.00 prize to the first boy who would fly a kite across the gorge and fasten the end of the kite string to trees on the other side. From this string, a traveler rope, from which the cables were later to be strung, was carried across the gorge. Then Ellet hoisted himself into a basket and had himself moved across on pulleys while the crowd below shrieked with excitement.

As fortune had it, Roebling had to relieve Ellet and actually did every bit of the construction of the Niagara Suspension Bridge. Ellet, who had constructed a footbridge for use as a temporary passage, argued with the bridge company about the toll on the little bridge. He walked off. Roebling was called in and took the position as a salaried engineer and was granted the stipulation that he could oversee every step of construction.

The Niagara railroad bridge had a span of 821 feet and was 245 feet above the foaming river. It had two decks: one for trains and one for carriages and pedestrians. It succeeded in reassuring the public that suspension bridges could work. The bridge stood until 1891, when it was replaced by a steel arch bridge which accommodates heavier trains, for the arch is a stiffer form.

It is believed that during the construction of the Niagara bridge, John Roebling's mind fixed upon the idea of constructing the impressive span to cross the East River. Visiting a cousin in Brooklyn in the winter of 1852, Roebling took the Atlantic Avenue-Fulton Street Ferry. Ice floes blocked the boat's passage and it took several hours to land. Roebling realized the advantages the bridge would bring to the growing city of Brooklyn, and the fame it would have, located in the most important city of the continent. The bridge, he believed, should cross the river at Blackwell's (later Welfare, now Roosevelt) Island. The idea was germinating in Roebling's mind during the construction of the Niagara span.

It was not until Roebling completed the Niagara Bridge that he could thoroughly study the problems and costs involved in creating a bridge across the East River. In 1857, already working on the suspension bridge between Covington and Cincinnati, Roebling wrote a letter to Abram S. Hewitt, later to become mayor of New York and a United States Representative, to say that such a bridge was possible, and would be highly beneficial to the two growing cities. Hewitt was impressed. The idea had been suggested before, but not by such an eminent engineer. Hewitt placed the letter before the public by having it printed in the *Journal of Commerce.* It aroused considerable support, especially from Brooklynites faced with the problems of winter commuting. However, the tensions of the impending Civil War and a minor economic crises overshadowed Roebling's scheme, and it was not until after the war ended that action was taken upon his suggestion.

Meanwhile, Roebling went to work on the Cincinnati-Covington Bridge. Ellet's bridge crossing the Ohio had proved the river could be spanned and Roebling had been hired to build a somewhat longer bridge across the Ohio.

The building of the Cincinnati-Covington Bridge took 10 years. It was complicated by political finagling, public apprehension, economic scares and the war itself. It was here Roebling learned the difficulties that politics and public sentiment can cause in the building of a public work.

As the war ended, Roebling, who was tired of city governments, approached some prominent Brooklyn businessmen with his plan. By now he had decided the bridge should link the city halls of the two cities and had a cost estimate of $4 million. He explained how the suspension type bridge would be perfect for New York harbor as the stone piers necessary for traditional bridges blocked the navigation of large vessels.

He succeeded in convincing William C. Kingsley, a leading Brooklyn contractor and businessman with political connections, and publisher of the very influential *Brooklyn Eagle,* that the bridge would serve the future interests of Brooklyn and was sound engineering.

One night in December, 1866, Kingsley and a friend paid a visit to the home of Henry Murphy, a State Senator who had formerly been the mayor of Brooklyn. They stayed up late into the night discussing the merits of the proposal and its political overtones. Murphy consented to introduce a bill into the state legislature which would enable a private company to build a toll bridge between the two cities. So it was in Brooklyn, the less developed, poorer city, that the first step was taken.

Murphy promptly introduced the bill. The winter of 1866-67 was unusually harsh, which aided in getting support for the bill's passage. A common reference to the difficulties in commuting from Brooklyn to Manhattan was that the time spent was greater than the time it took to travel the 142 miles by train from Albany to New York. Newspapers took up the popular appeal. In April, a bill "to incorporate the New York Bridge Company for the purpose of constructing and maintaining a bridge across the East River" was passed.

Under the enabling act, the city of Brooklyn had to subscribe for $3 million of the capital stock, while New York paid only half that, for it was expected that Brooklyn would derive greater profit from the new bridge. Basically, Manhattanites were apathetic. They were to grow absolutely hostile as the expenses for the bridge continued to climb beyond all expectations and the bridge company, whose stock was partially owned by private citizens, was suspected of scandal.

Thirty-nine incorporators were named in the bill. They elected Senator Murphy as president. Kingsley held a large block of the stock, as did Murphy. This commingling of public and private funds to finance what was basically a public project had a number of precedents, but in this case was to lead to a major investigation. The company was to be permitted to fix toll rates for pedestrians or any kind of vehicles, receiving a profit of no more than 15% per year. Although the two cities subscribed to most of the stock, they had no voice in the planning or location of the bridge.

Twin Gothic-style arches pierce the bridge's granite towers

PHOTO BY PETER ROSE

The company unanimously chose Roebling to be chief engineer. He immediately came to New York and by May, 1867, had assumed the position he had dreamed of for so many years, at a salary of $8,000 per year. He sent his son Washington, who had earned an engineering degree at Rensselaer Polytechnic Institute in New York, to Europe to study newly developed methods of sinking caisson foundations. Young Roebling, who had built bridges for the Union Army during the war and had helped his father on the Cincinnati Bridge, also visited the Krupp and Essen works in Germany to study a new structural material—steel.

It was imperative to Roebling that he be given free rein to construct his bridge. In preparing a preliminary plan for the bridge company, he was cautious in a way that showed his understanding of the fickleness of politicians and the public. He not only supplied statistical data and descriptions of the bridge but also dwelled on its usefulness, soundness, and its potential for revenue, which he hoped would enthuse the public-spirited and mollify skeptics.

He personally presented this plan to the company three months later, assuring his audience that the building of such a long structure did not detract from its stability, but that "Any span inside of 3000 feet is practicable . . . A span of 1600 feet or more can be made virtually as safe and as strong in proportion as a span of 100 feet. The larger span is a question simply of weight." In addition, to guard against movement that could possibly be caused by winds and storms, Roebling described his system of iron trusses which run the length of the suspended floor from anchor wall to anchor wall, which would possess ample stiffness even in the greatest emergency. He also emphasized the usefulness of inclined stays running diagonally downward from the towers—those fine tracery lines that give the Brooklyn Bridge its ethereal quality. These wires, he contended, used

COURTESY OF THE NEW YORK MUNICIPAL REFERENCE LIBRARY
This Christmas card was drawn by engineer David Steinman.

in conjunction with the main cables would make the bridge so safe that "if the cables were removed, the Bridge would sink in the centre, but would not fall . . ." Roebling noted too that the use of steel for the first time in a suspension bridge would ensure a strength stability never before attained.

As far as the usefulness of the bridge was concerned, Roebling foresaw 40 million people crossing annually. He underestimated, for it is believed that as many as 150 million passengers and pedestrians cross the Brooklyn Bridge each year.

The bridge he described was to have five lanes. The two outer lanes were to serve horse drawn carriages. The middle lanes were for a cable train with terminals at either end, for which passengers would be charged five cents a ride. In this way, Roebling claimed, the bridge would pay for itself in railway passenger fares, obviating the need for vehicular tolls on the roadway. The fifth lane Roebling described as "an elevated promenade which will allow people of leisure old and young to stroll over the bridge on fine days. I need not state that in a crowded commercial city such a promenade will be of incalculable value."

This wide center walkway, elevated 18 feet above traffic, is still a delight to those with time for a stroll. It affords an unrivaled view of the Lower New York skyline, and unlike walks on other traffic bridges, makes the pedestrian feel he is surrounded by a glittering cat's cradle.

Appealing to businessmen, Roebling pointed out that two large fireproof spaces in the arches could house treasury vaults or serve as storage for private entrepreneurs, bringing in more revenue for the bridge company. For many years these underground cellars stocked the choicest wines in New York. The Eighteenth Amendment halted this. On the bridge's 60th anniversary Meyer Berger of the *New York Times* noted that 14 bustling industrial concerns paid the city an average of around $27,000 a year for work space and storage inside the span's landside arches. (Recently, an architectural plan has been presented for converting some of the arches into a gallery and center for artists). Roebling also assured them that skyrocketing land values in Brooklyn would pay for the cost of the bridge within three years.

In addition, for those who still felt the bridge an unnecessary, wasteful contraption, Roebling predicted that the natural population growth of the two cities, of Brooklyn in particular, would necessitate the building of still more bridges over the East River, and possibly tunnels as well. Prophetically, he suggested bridges crossing the river to the Williamsburg section of Brooklyn, and another crossing over Blackwell's Island. In fact these two bridges, the Williamsburg and Queensboro, were the next to span the river, built at the turn of the century.

Many were still unconvinced. The proposed weight the bridge was to carry—18,700 tons—astounded many, including Mayor Kalbfleisch of Brooklyn and Horace Greeley, publisher of the influential *New York Tribune.*

However, the current mayor of New York—"Boss" Tweed—was not about to listen to these doubters. In the bridge project Tweed saw another potential source of lucrative graft. As a subsequent investigation was to show, Tweed, encouraged by $65,000 in cash and the promise of stock options in the Bridge Company, pressured the City Council of New York to consent to New York's financial participation in the Bridge Company.

Still, Roebling wasn't satisfied with his reception. He proposed setting up a panel of engineers made up of competitors to judge the thoroughness, safety, economy and practicability of the design. In addition, the U. S. Secretary of War appointed three army engineers to ascertain whether the bridge obstructed navigation.

Both groups unanimously approved the bridge by June 1869. The army engineers fixed the height of the roadway at 135 feet, the clearance measure still used today. In fact, this was later to become the maximum height of U. S. Navy vessels as they had to pass under the bridge to the nearby Brooklyn Navy Yard.

Riding on the crest of his greatest success, one week after all opposition was overcome Roebling had an accident from which he died three weeks later, never to see one bit of the physical construction of his last design. While standing on some piles on the Brooklyn shore, pinpointing the exact location of the Brooklyn tower, Roebling was so engrossed in his labors that he did not see a ferry boat enter its slip clumsily. As it crashed against the slip, the piles fell and Roebling's leg was crushed. The leg had to be amputated. Roebling believed in faith healing and cold water treatment and refused any medical aid. After three weeks of severe pain, conscious to the end, Roebling died, leaving his son to fulfill the task.

John Roebling had left voluminous notes as to the design of the bridge, but Washington Roebling needed an equal amount of ingenuity and concentration to overcome structural dilemmas.

The first major problem came during the sinking of the pneumatic caissons. These are huge, airtight, bottomless boxes or cylinders in which men work in an atmosphere of compressed air doing excavation under the river bed. The men, called "sand hogs," clear off layers of material to get the caisson to sink gradually while at the same time it is built up at the top to keep the walls above water level. The caissons must eventually reach solid rock in order to support the weight of the bridge towers.

It was dangerous work. In the early stages

Above: Early photograph of the Brooklyn Bridge showing original transportation plan. Horse drawn carriages used the outermost lanes, trolleys the center lanes and elevated trains used the innermost lanes. Below: Diagram shows clearance above the East River.

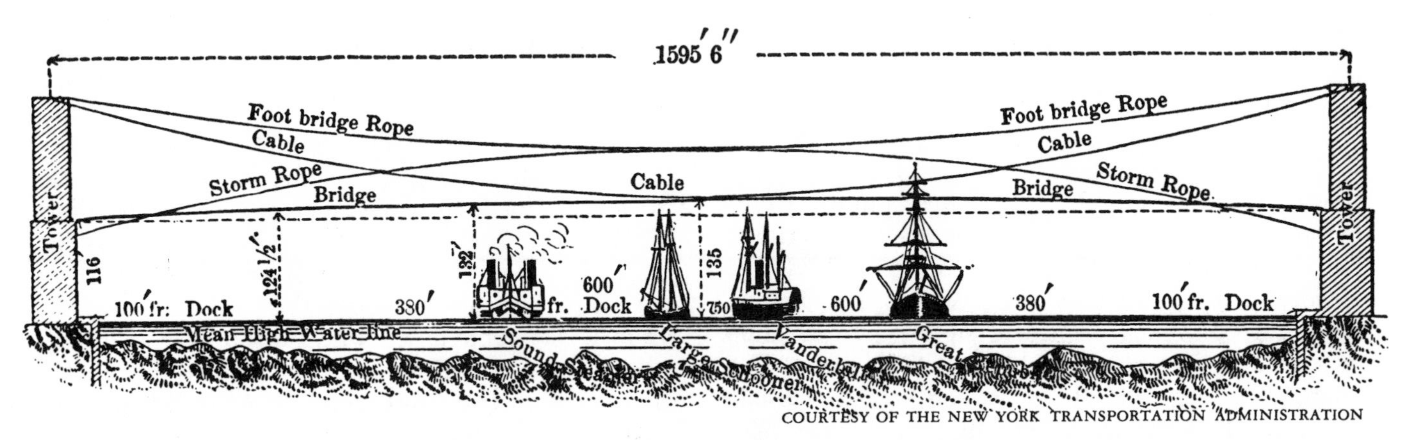

COURTESY OF THE NEW YORK TRANSPORTATION ADMINISTRATION

COURTESY OF THE NEW YORK TRANSPORTATION ADMINISTRATION
The Brooklyn caisson was the first sunk into the riverbed. The bottomless wooden box was 168 feet long and 102 feet wide.

"blowouts" often occurred: a gulp of compressed air would rush out of the caisson, permitting water to enter the chamber. This would cause a great flume of water and debris to shoot up into the air through the various entrances to the caisson and bombard the men above who were working on the masonry arches. The work chambers were lit with calcium or gas lamps which gave them an eerie quality and were also dangerous, as the caissons were primarily constructed of wood. The noise of the river currents and passing boats reverberated through the caissons. It was difficult for workers to speak, though for some unknown reason, time seemed to pass very quickly while working in compressed air. One engineer likened the atmosphere to Dante's Inferno and the men's labors to those of Sisyphus. However, the worst danger was "the bends" or caisson disease, which can cause paralysis and even death. All known safety precautions were taken. At first men worked in three shifts of eight hours each, but shifts were shortened to two hours when the caissons got deeper and the air pressure had to be increased. The men got hot coffee and were examined when they came out of the compressed air atmosphere. But the exact cause of the disease had not yet been determined. Six years later physicians were to discover that it was the rate of decompression which was the critical factor in releasing nitrogen bubbles from solution in the blood and tissues, causing cramps, paralysis and death.

The rate of the Brooklyn caisson's descent was slowed to six inches a day when huge boulders were found embedded in the soil. The question of blasting the rocks came up. The effects of using blasting powder in compressed air was unknown, but Washington Roebling experimented by shooting a gun in various parts of the subaqueous chamber. He found there were no ill effects, and gradually used larger and larger charges. Once blasting powder was used, the caisson sank at the rate of between 12 and 18 inches a day, until the Brooklyn foundation reached a level of 90 feet below the river's bed. It took one and a half years to sink the caisson, and Washington Roebling spent much of his time personally supervising the sand hogs. He was also invariably present when any accident or fire occurred, enduring hours of work far beyond what he would expect of his workmen.

In 1870 there was a major fire in the Brooklyn caisson. A gas lamp apparently was left too close to the ceiling of the working chamber, which caught fire. Roebling remained in the chamber for 7 hours, at which point he collapsed and had to be carried out. Four hours later he was back working. Burning embers were discovered in an enclosed layer of timbers, so Roebling had the entire caisson flooded. It took over a million gallons to fill the enormous box. After 2½ days carpenters had to be called in to reline the chamber. Eventually, cement was injected throughout the burned out roof.

The New York caisson presented even more difficulties, as an irregular surface of gneiss rock requiring heavy blasting was uncovered under the surface of the riverbed. Then boiling quicksand was discovered. At 70 feet a level of cemented gravel was reached. This gravel was so hard that it was impossible to drive an iron rod through it without battering the rod to pieces. It was decided to let the caisson settle at this point. Despite the fact that this caisson is not as deep as the Brooklyn one, it took three years for it to be completed. During that time Washington Roebling, or Colonel Roebling, as he was called, worked along with his men, giving them confidence and encouraging them. Although only six men had come down with caisson disease during the sinking of the Brooklyn caisson, there were 110 cases on the New York side, 70 of which had to receive medical treatment. Three men died of it. In

Above: Portrait shows Emily Warren Roebling, the wife of Col. Washington Roebling. She was educated and confident enough to relay her husband's plans to workers and managers. Below: Diagram of cable spinning.

the early summer of 1872 Colonel Roebling had to be carried out of the caisson nearly insensible. Physicians expected him to die. Although he took long cures in Europe, he remained a semi-invalid to the end of his life. Fearing death, he began in a frenzy to write out minute plans and directions for completing the masonry work, building the anchorages, stringing cables and suspending the wooden walks from which the suspenders and roadway were to be constructed. His wife, Emily Warren Roebling, took over the supervision of the project as Colonel Roebling continued to watch the proceedings through binoculars from his Brooklyn Heights apartment. During the remaining 10 years the bridge was under construction, Roebling became progressively worse. It was falsely rumored that he had to communicate instructions to Emily via Morse Code.

It took five years in all to build the two main towers. Meanwhile the anchorages, which are enormous masonry structures in which the cables are embedded, were constructed. The site of the New York anchorage is an historic one. Bounded by Cherry, Dover, Roosevelt and Water Streets, it is located where George and Martha Washington lived from 1789 to 1790 when New York was the capital of the United States.

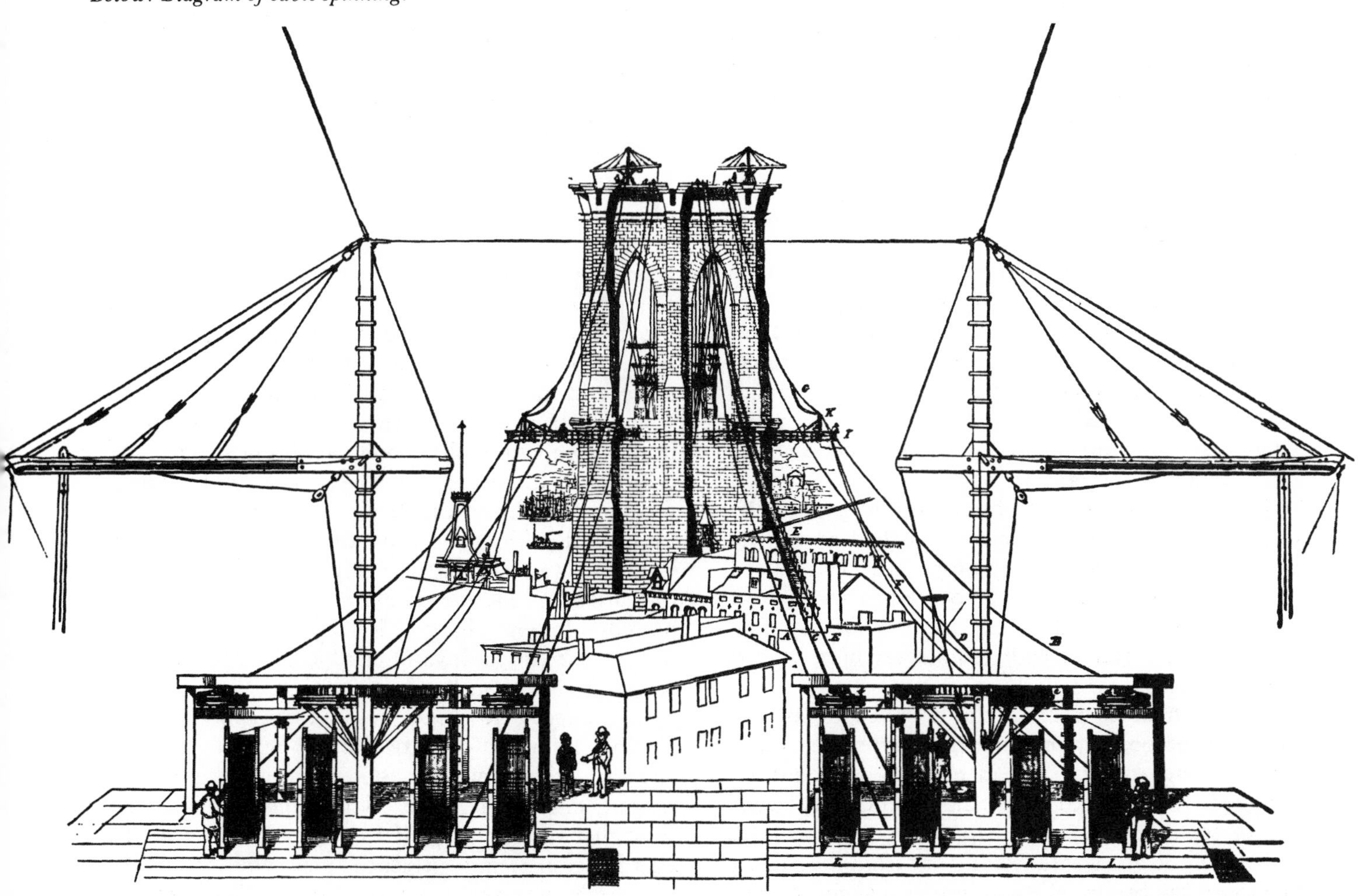

COURTESY OF THE NEW YORK TRANSPORTATION ADMINISTRATION

In 1876, towers and anchorages were completed, and the enormous parabolic cables had to be woven. All wires were to be made of steel with a tested strength of 160,000 pounds per square inch, and bids from wire companies both in America and Europe were entertained. Washington Roebling, adhering to a bridge company rule which forbade a contract going to any company in which an officer or engineer of the bridge company had an interest, resigned his interest in John A. Roebling's Sons. He felt that this firm, already the biggest wire rope company in America, was the only reliable supplier. A controversy arose as to whether Bessemer or crucible steel should be used in the cables, and bids were submitted for both types. While John A. Roebling's Sons submitted the lowest bid for Bessemer steel, the bridge company, directed by a report written by Abram Hewitt, decided to award the contract to J. Lloyd Haigh Co., New York, the lowest bidder for crucible steel wires.

Although Hewitt was a friend of Haigh, it had long been believed that he received no financial kickback for getting the contract for him. At worst it seemed to be a political favor; at best an error in judgment. Hewitt's reputation during his own lifetime was that of a trustworthy, crusading liberal. In fact, after Mayor Tweed was ousted in 1871, Hewitt had been appointed to investigate the management of the bridge company, to rid the public of any suspicion of corruption in the building of the bridge. Besides the Tweed escapade, Hewitt reported, Kingsley had been overpaid for some shares in the bridge company, but all else was on the up and up. Subsequent research into the diaries of Washington Roebling, however, showed that Hewitt may not have been so selfless a politician.

According to Washington Roebling, Hewitt held the mortgage on Haigh's industrial property, and that for as long as Haigh held the contract, Hewitt received 10% of Haigh's payments from the bridge. The engineer also noted that much of the wire Haigh sent was in fact made of the cheaper Bessemer steel, which gave him an additional profit. Roebling did not take any action at that point.

In 1878, however, Roebling had to report the discovery that defective wires, which had previously been rejected by official inspectors, had been passed off as approved wire. This discovery had taken a bit of detective work, and no one was sure exactly how much of the defective wire had already been woven into the cables.

A workman inspecting the wire ropes at the Haigh plant noticed that the pile of rejected cable was growing smaller, not increasing, as would be expected. He reported this. Another man was sent out to see just what was happening. Following a wagonload of good wire setting out from the Haigh factory to the bridge, the man found that the wagon was driven into a second building where it was unloaded and replaced by a load of rejected wire. The good wire was taken back to the first plant to be inspected again.

Roebling had to decide whether to completely unwind the cables in order to discard the substandard wire. The cables are composed of hundreds of wires laid parallel to form 19 strands. The strands in turn are wrapped together and make up the cable. Too much work and expense and time would be involved in dis-assembling the cables, so Roebling added good wires to the cables to make up for any possible deficiency in strength. These had to be supplied at Haigh's own expense. For this reason, the number of strands woven into the Brooklyn Bridge's cables is uneven.

Although this safety hazard was overcome without additional expense, the scandal caused additional debate about the soundness of the bridge. The last wire for the cables was run across on October 5, 1878, and already the absolute limit set by the legislature of $13.5 million for construction had been exceeded. A taxpayer's suit had been filed. New York City Comptroller John Kelly claimed the bridge was "a scheme, by the machinations of a private corporation to secure the expenditure and control of public funds." A society called the New York Council of Reform campaigned against the bridge, noting that the bridge was "wholly experimental," that five of the largest suspension bridges in Europe and America had recently been destroyed by weather, and that the East River Bridge would collapse in the first big gale.

Litigation dragged on for months, and by November 1878 construction work had to cease for lack of funds.

Six months later new stipulations were attached to funds appropriated by the state legislature. Construction could begin again, but the bridge now had to provide for the weight of heavier trains. Roebling had to design additional stiffening trusses, which again cost more money, and from 1879 until 1883 when the bridge was finished, relations between the bridge company and Colonel Roebling deteriorated drastically. In fact, a mere year before the bridge was completed, the Bridge Company took measures to have Roebling relieved of his position. First Roebling was accused of being crooked, selling information on bids and taking graft. When he cleared himself of these charges, it was insinuated that he was a half-crazed invalid, and could not take the responsibility of such a huge project.

Washington Roebling refused to appear at the hearing. In his notes, Roebling comments that the directors who now requested his presence at this judgment were the same "who joined with the request that I absent myself from any meeting of the Board because my presence may embarrass Mr. Kingsley's proposed operations of putting a couple

Dignitaries pose on catwalk before cables were spun. COURTESY OF THE BROOKLYN PUBLIC LIBRARY BROOKLYN COLLECTION

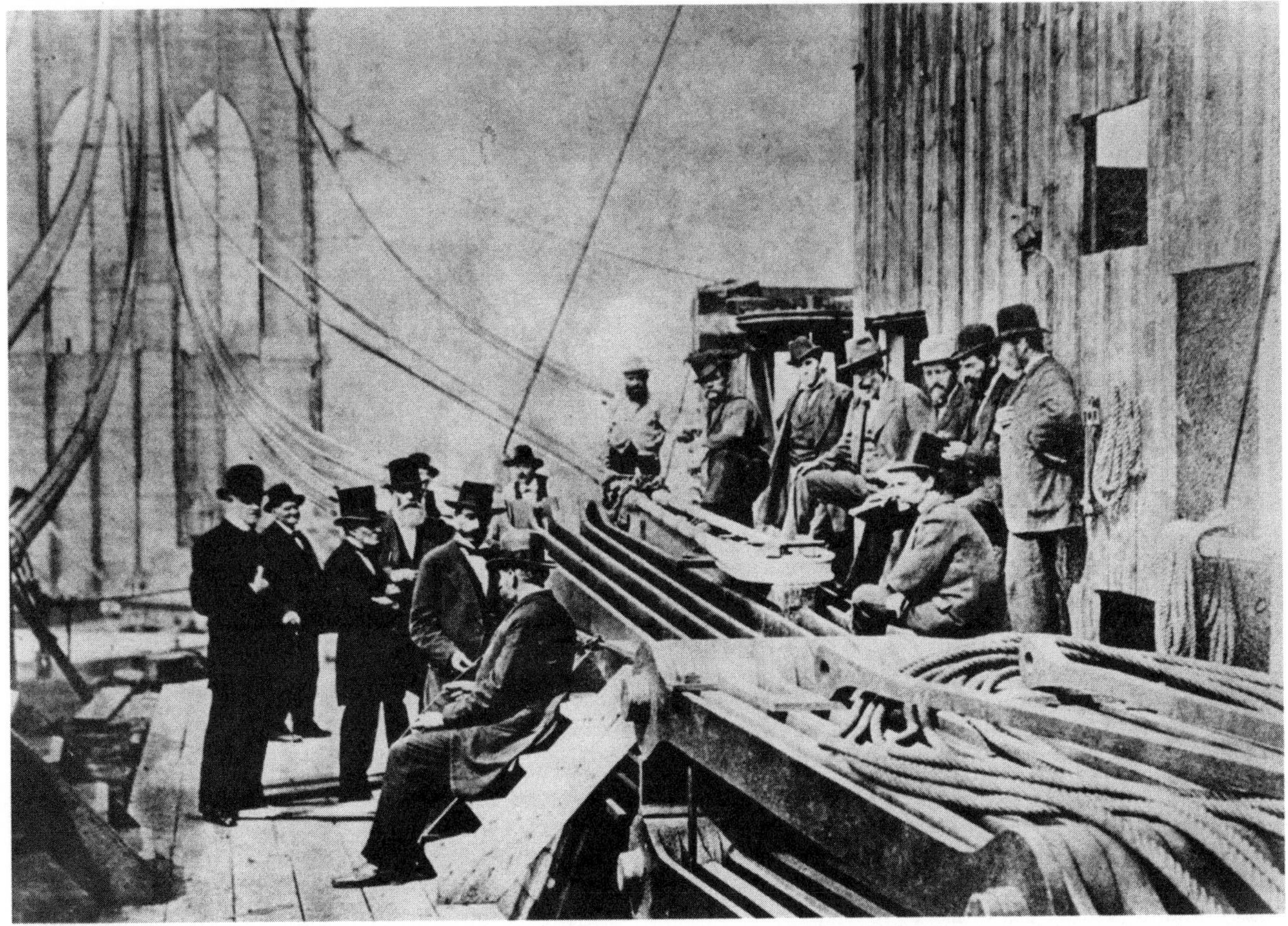

Stockholders and engineers inspect cable spinning. COURTESY OF THE NEW YORK TRANSPORTATION ADMINISTRATION

PHOTO BY PETER ROSE

of millions in his pocket, millions which have not yet reached their destination." Emily Roebling, who, from detailed instruction from her husband had been actually overseeing the construction, appeared at the meeting and requested that her husband be permitted to continue in his position as Chief Engineer. The board complied with her wishes, but the bitterness remained throughout the completion of the bridge.

The Opening Ceremonies took place on May 24, 1883. New York's Irish population threatened to boycott the celebration because it was on the same day as Queen Victoria's birthday. Bridge company officials denied that there was any connection, and refused to change the date of the proceedings because elaborate preparations had already been made. Whether the Irish participated or not, the festivities and ceremonies were described by newspapers as the most grandiose New York had ever seen.

Schools and businesses were closed for the day in Brooklyn, while New York had a half-day holiday. Both cities were decorated with banners and bunting, pennants and flags. Fireworks displays and parades lasted far into the night, and the fire department sent out boats to light up the river. A sign in a Brooklyn shop window read "Babylon had her hanging gardens, Egypt her pyramids, Athens her Acropolis, Rome her Athenaeum; so Brooklyn has her bridge."

President Chester A. Arthur and his entire Cabinet attended the official ceremonies, which took place at the Brooklyn terminus. Grover Cleveland, governor of New York, was there, as were Mayor Seth Low of Brooklyn and Mayor Franklin Edison of New York. Thousands watched and cheered the procession of dignitaries. When they arrived at the bridge, the guns from the forts in New York harbor and from ships of the United States fleet anchored near the bridge saluted them, and the chimes from Lower Manhattan's other Gothic monument—Trinity Church—joined the clamor.

After the march over the bridge and the speeches at the terminus, the officials walked the short way to Washington Roebling's home, where they congratulated him. Whether it was his illness which forced him to stay away from the celebration or his feud with the bridge company is unknown.

At the opening ceremonies, William Kingsley, President of the board of trustees of the Bridge Company, formally presented the bridge to the cities of New York and Brooklyn. The two mayors made the acceptance for the people and then Abram Hewitt delivered the major oration. This speech was apparently a favorite of Hewitt's, as he later had it reprinted as a pamphlet when he ran for mayor of New York.

Hewitt also compared the bridge to the pyramids. He called the bridge "a new glory to humanity" and called on his audience to appreciate it as a monument to the possibilities inherent in the democratic spirit combined with technological progress.

But in a more concrete tone, Hewitt examined the political possibilities of the bridge. Although there was now a physical link between the two cities, Hewitt warned there was no need to change political relationships.

"While we rejoice together at the new bond between New York and Brooklyn we ought to rejoice the more that it destroys none of the conditions which permit each city to govern itself but rather urges them to a generous rivalry in perfecting each its own government, recognizing the truth that there is no true liberty without law and that eternal vigilance, which is the only safeguard of liberty, can be best exercised within limited areas."

Although Hewitt seemed to have a strong conviction that the two cities should remain separate entities, he pointed to one serious effect of the bridge, for it is quite probable the construction of the Brooklyn Bridge gave the final impetus to the politics of consolidation.

"That Brooklyn," Hewitt went on to predict, "will gain in numbers and in wealth with accelerated speed is a foregone conclusion. Whether this gain shall in any wise be at the expense of New York is a matter in regard to which the great metropolis does not concern herself."

Turning from the political, Reverend Richard Salter Storrs, a Brooklyn clergyman with a gift for rhetoric—both celestial and architectural—rhapsodized on the beauty and meaning of the bridge. He compared the bridge to the Arch of Triumph and the Brandenburg Gate, calling it a silver band which stretched over the silver streak of the divisive East River. Storrs envisaged the bridge as a monument to peace, a peace that was felt between these two cities willing to interlink their destinies.

"This structure will stand, we fondly trust, for generations to come, even for centuries, while metal and granite retain their coherence; not only emitting when the wind surges or plays through its network, that aerial music of which it is the mighty harp, but representing to every eye the manifold bonds of interest and affection, of sympathy and purpose, of common political faith and hope, over and from whose mightier chords shall rise the living and unmatched harmonies of Continental gladness and praise."

Not everyone agreed with Storrs on what he called the bridge's "harmonious proportions and dainty elegance." The first architectural critic to comment on the bridge was Montgomery Schuyler, who voiced

PHOTO BY PETER ROSE

The cables pass over saddles within the towers. Montgomery Schuyler, an architectural critic, charged Roebling with non-functional design for the towers.

his opinion for *Harper's Weekly Magazine* in April, 1883. Schuyler praised the bridge as a convenience and a piece of skillful engineering but was adamantly critical of the towers, calling them unfunctional.

> "If the structure had been architecturally designed, these (structural processes) would have been emphasized at every point and in every way. The great towers, so-called, designed merely to hold up cables, would have performed that function and the stability of the piers, loaded as they are by cables, would very possibly have been assured, even if they had been completely detached from one another."

He indicates that the bridge would have been better designed if there had been four separate towers rather than the two arches.

> "There are probably few arches in the world, certainly there can be none outside of works of modern engineering, of anything like the span, height, thickness and conspicuousness of those in the bridge towers, which are so little effective. Like the brute mass above them, they are impressive only by magnitude."

Several years later Schuyler further complained in an article in *Architectural Record,* that the towers should have been modeled with saddle-backed roofs, so they could assert their role as cable-holders instead of concealing the cables so that they seem "imbedded in the tower on each side and there to cease and determine, instead of being a necessary link in a continuous and mobile chain."

Schuyler considered himself a modernist and a functionalist, but Frank Lloyd Wright, the undisputed monarch of modern architecture, praised John Roebling as one of the few builders who influenced him.

Lewis Mumford and Talbot Hamlin, two eminent architectural critics, admired Roebling's towers in particular. Mumford, in the *Brown Decades,* praised the towers as the high mark of American architecture of the period between the designs of the Washington Monument and the late phase of (Henry Hobson) Richardson. "In this structure," he wrote, "the architecture of the past, massive and protective meets the architecture of the future, light, aerial, open to sunlight, an architecture of voids rather than solids." Hamlin, who taught architecture at Columbia University, thought the towers gave an "impression of immense power" similar to that of the court of Thebes, the Roman Colosseum and the Cathedrals at Rheims and Westminster. The majesty that these two critics saw in Roebling's towers gives a clue to the reason the bridge has figured so largely in the works of so many of America's painters and writers.

Walt Whitman wrote in *Specimen Days* of the "grand obelisk-like towers on either side, in haze, yet plainly defined, giant brothers, twain, throwing free graceful interlinking loops high across the tumultuous current below."

To Hart Crane scientific progress was united with poetry in the symbol of the bridge. The particular angle at which the poet could view the bridge through his window overwhelmed Crane. Critics say he was unaware he was occupying Washington Roebling's former room.

American authors were not the only ones who praised this bridge. Vladimir Mayakovsky, the Soviet poet, saw the Brooklyn Bridge in his travels in the 1930's and dedicated a poem to it. He, too, celebrates it as a monument to modern civilization. A contemporary Russian poet, Andrei Vozhnezenski, reverently praises the Brooklyn Bridge in his poem "Kennedy Airport." Vozhnezenski's appreciation is not completely aesthetic, one would imagine, as Andrei Vozhnezenski holds a degree in architecture.

Mumford summarized the impact of the Brooklyn Bridge in his book *Sticks and Stones.* In 1929 he wrote,

COURTESY OF THE AUGUST ROCHLITZ COLLECTION

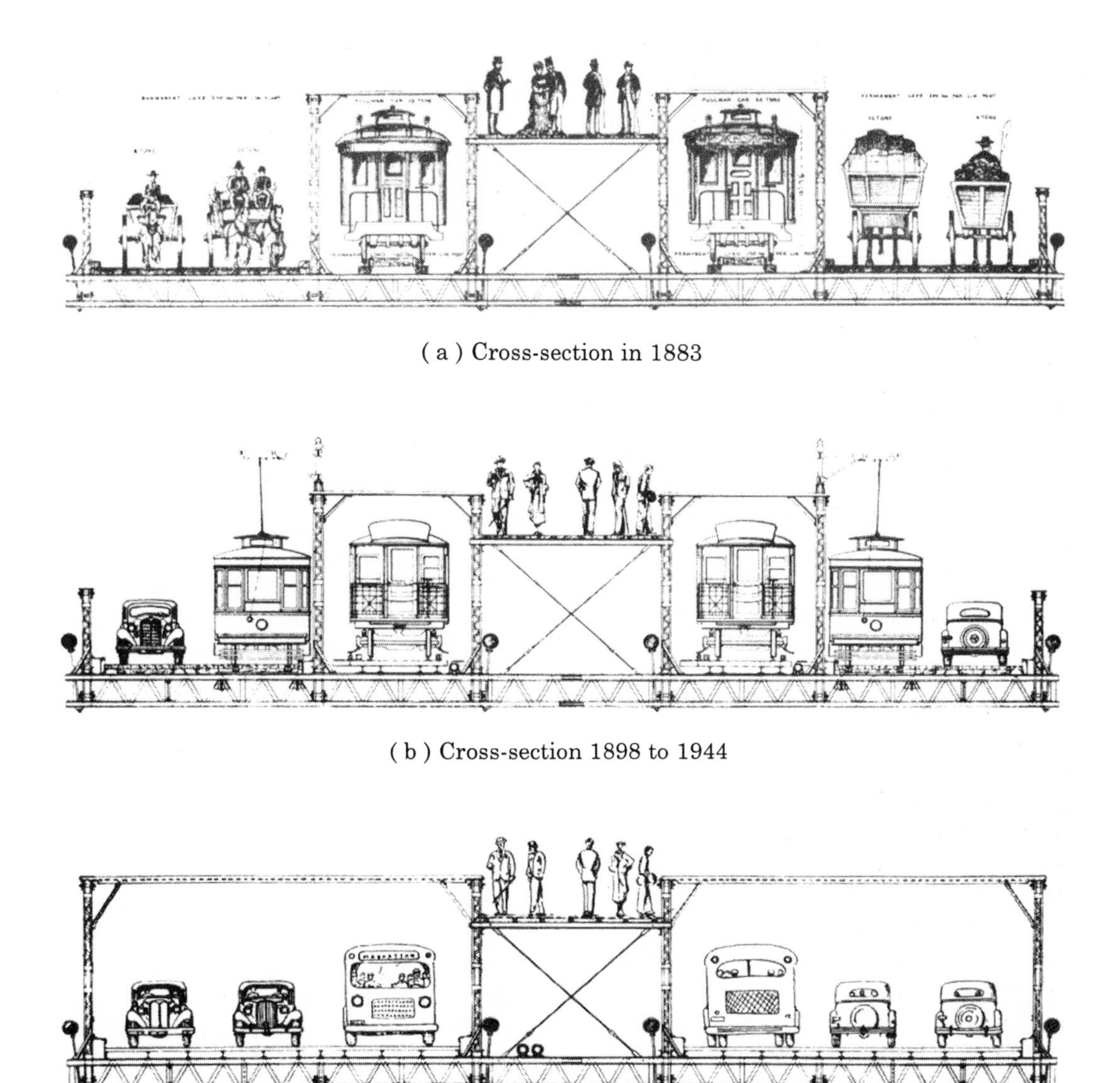

(a) Cross-section in 1883

(b) Cross-section 1898 to 1944

(c) Proposed cross-section

COURTESY OF THE NEW YORK TRANSPORTATION ADMINISTRATION

"Beyond any other aspect of New York, I think, the Brooklyn Bridge has been a source of joy and inspiration to the artists . . . that the age had just cause for pride in—its skill in handling iron, its personal heroism in the face of dangerous industrial processes, its willingness to attempt the untried and the impossible—came to a head in the Brooklyn Bridge."

Thus, the immigrant painter Joseph Stella found the Brooklyn Bridge "a joyful, daring endeavor of the American civilization." Both he and watercolorist John Marin painted numerous versions of the bridge, romantic as well as structural.

In *You Can't Go Home Again,* Thomas Wolfe, who attempted to weave together the complex aspects of America and all of civilization, writes of his hero walking across the Brooklyn Bridge after forays into Manhattan. In Arthur Miller's *View from the Bridge* and Maxwell Anderson's *Winterset* the Brooklyn Bridge is used as a backdrop for the unfolding of lives among the poor and immigrant groups who lived near the docks. In fact, during the depression, large shanty towns were set up under the bridge.

These gloomier perspectives of the bridge are not without precedents in the bridge's history. The Brooklyn Bridge has had more than its share of deaths and tragedy. During the building itself, 20 men were killed, including John Roebling. A panic which killed 12 more occurred a week after the bridge opened. The bridge was thrown open free of charge to pedestrians in honor of Memorial Day. The traffic was very dense, owing to the novelty of the bridge. One woman who was walking up the steps on the Manhattan side tripped, and her female companion screamed. The scream triggered off a rumor among those walking on the promenade that the bridge was falling. Those who were entering on the Manhattan side continued to climb up while those already on the bridge tried to jam their way down the narrow stairway. In the resulting crush 12 were killed and 35 seriously wounded.

The bridge also seems to have a preternatural attraction for those souls who long for some kind of glory or who wish to do away with themselves. The most famous of the former is Steve Brodie, whose name, in connection with jumping off the Brooklyn Bridge, has become an Americanism. In fact, an elaborate, expensive David Merrick musical comedy written about Steve Brodie enjoyed one performance in New York.

On July 23, 1886, Steve Brodie, a hanger-on in the theatre district, which was then centered around the Bowery, was found in the East River under the Brooklyn Bridge. Although he received much publicity, he could not prove he had jumped, since there were no accredited witnesses to attest to his feat. Thus, if he did jump, he jumped in vain, and if he did not jump he has achieved much mistaken acclaim. He cashed in on the fame, and operated a bar on the Bowery, which was frequented by boxers and celebrities as well as sightseers. "To do a Brodie" has been given two meanings. It means to commit suicide by jumping off a bridge and it also means a failure, mistake or blunder.

Brodie was preceded in his leap for fame by Robert Emmett Odlum, a Washington swimming instructor. Odlum had announced that he would make his leap as early as 1882, when he made a $200 bet that he would jump from the bridge into the East River. At the time he was unable to obtain a pass onto the unfinished structure. He sneaked on and started undressing, but was forcibly removed by bridge employees. In 1884 he again attempted the jump. Police had been warned about it, and they had been watching for him for several days. On May 20, 1884, a cab which seemed to be making "unseemly haste" aroused the suspicion of a tolltaker. Reporting it to the police, he claimed that there were two gentlemen, one of whom seemed to be hiding the other. The police stopped this cab when it got near the railing at the middle of the bridge. It turned out, however, that this was a decoy cab, and Odlum was riding three vehicles behind in an inconspicuous wagon.

Odlum succeeded in jumping, but his body hit the water at a peculiar angle. His friends, waiting in a boat below, screamed for help and eventually managed to rescue him from the strong East River current. He was unconscious, and died less than half an hour later from internal hemorrhages.

Odlum's leap was well publicized and the succeeding weeks saw a large number of attempted suicides from the Brooklyn Bridge. One destitute German immigrant, who was nabbed by the police before jumping, aroused so much public sympathy that he received many gifts and offers of jobs. This, too, was publicized and may have contributed to others attempting to "commit suicide."

Another Americanism which involves the Brooklyn Bridge is the phrase, "You could sell him the Brooklyn Bridge," which, of course, labels a person as a country bumpkin. The origin of this phrase is uncertain. Some think it is due to the changes in ownership of stock from the Bridge Company to the two cities, and a third changeover when the two cities became one. Others attribute it to the fact that the bridge was for many years the most prominent, most expensive structure and buying it was akin to buying a pyramid.

In 1944 the El trains which ran to terminals on either side of the bridge were eliminated. The terminals were architecturally interesting structures, but fouled up traffic, and were also razed. David Steinman was hired to remove the trusses which had stood between the train tracks and the roadway. At that time he also strengthened the outside trusses which had formerly carried 10% of the weight of the suspended structure, and now carry 50% of it. At the 75th anniversary celebration and second opening of the bridge. Steinman said "I still find it aesthetically the most satisfactory bridge in the world."

COURTESY OF THE AUGUST ROCHLITZ COLLECTION

Above, the elevated terminal at the Manhattan end of the bridge. Right, the view from the back seat of a trolley car.

PHOTO BY PETER ROSE

WILLIAMSBURG BRIDGE

UPRIVER USURPER

"The aim of the Greeks in their works was beauty. The object wrought in our time is utility, and the Williamsburg Bridge is the very temple, the highest achievement of the utilitarian," wrote the editor of the *Brooklyn Daily Times* the day the first elevated train went into service on the bridge in 1905.

Utilitarian philosophers may not reject beauty, but it is true that the Williamsburg Bridge has never been known for its loveliness. It supplanted the Brooklyn Bridge as the longest suspension bridge in the world. It was the first suspension bridge to be constructed with towers entirely of steel. It was the world's strongest suspension bridge. It provided a welcome outlet for the Jewish immigrants trapped in the overcrowded tenements of the Lower East Side of Manhattan.

If the bridge is at all colorful, it is due to its proximity to those narrow streets full of pushcarts, old world shops and newly-arrived Russian Jews with their exotic customs and atmosphere. If the bridge is immortalized it will be in the stories of sentimental chroniclers of Lower East Side Jewry, like Harry Golden. One can still hear former residents of the neighborhood reminisce about evening walks on the bridge in order to escape the steamy tenements.

The bridge's 350 foot tall towers dominate the

adjacent neighborhoods like oil rigs among Oklahoma shanties. They are stark, skeletal. The trusses are ponderous. It would seem the designer, Leffert Lefferts Buck, had little interest in designing an architecturally harmonious structure. The use of steel was relatively new in 1896, when Buck planned the bridge, and this may be a factor in its clumsy appearance. The towers seem more closely related to the rigid, wrought iron designs of the previous period than the steel towers that were to come. In fact, the concept of the rigid steel tower was to be discarded by the time the Manhattan Bridge was designed in 1903. But Buck's towers could quite possibly have been influenced by the works of Alexandre Gustav Eiffel, a French bridgebuilder most famous for his tower for the Paris Exhibition, which had received wide acclaim in 1889. In fact, it is possible that Buck met Eiffel, as both engineers were involved in the building of bridges for South American railways in the 1870's. It should be noted that steel was available to Eiffel when he was building the tower, but he avoided using it because he felt his ideas might not translate into the new medium.

The Eiffel Tower had a few vocal critics, such as Guy de Maupassant. The Williamsburg Bridge seems to have been criticized by everyone. Even David Steinman, who at the age of 14, (in 1900) had obtained a pass from the city to walk around on the temporary roadway while the suspenders were being worked on, called the design ungainly and clumsy.

Scientific American, shortly before the completion of the bridge, harshly criticized its lack of good design.

> "Considered from the aesthetic standpoint, the New East River Bridge is destined always to suffer by comparison with its near neighbor the New York & Brooklyn Suspension Bridge. Whatever criticism has been made against the constructive features of the latter structure, it has always been conceded to be an extremely graceful and well-balanced design. It is possible that, were it not in existence, we would not hear so many strictures upon the manifest want of beauty in the later and larger East River Bridge, which is destined to be popular more on account of its size and usefulness than for its graceful lines. As a matter of fact, the East River Bridge is an engineer's bridge pure and simple. The eye may range from anchorage to anchorage and from pier to finial of the tower without finding a single detail which suggests controlling motive, either in its design or fashioning other than bald utility."

The bridge had been contracted and designed before Brooklyn and New York had united, so there had been no single official body of municipal engineers to pass or reject the design. Members of the New East River Bridge Commission were chosen equally by the mayors of Brooklyn and New York. When the two cities, as well as Queens and Staten Island joined to form Greater New York in 1898, there was a delay while the enlarged city chose new commissioners. The second commission permitted Buck to go on with his work after some controversy. Salem H. Nales, a member of the first East River Bridge Commission, warned that the designers of the bridge were disregarding the aesthetic elements of the problem and urged that steps be taken to beautify the towers. When reform candidate Seth Low was elected mayor in 1902, he appointed Gustav Lindenthal bridge commissioner. Lindenthal, who rejected any schism between engineering and art, said at a meeting of the Municipal Art Society that the towers of the New East River Bridge were "the ugliest possible." In order to get control of the Williamsburg Bridge, Lindenthal kicked Buck upstairs, making him a consulting engineer, relieving him of the position of Chief Engineer, and hired Henry Hornbostel, an architect, to attempt some kind of aesthetic rehabilitation. Lindenthal and Hornbostel were to collaborate on two of New York's greatest bridges—the Queensboro and the Hell Gate Bridge.

Montgomery Schuyler, the architectural critic who was firmly opposed to the lack of functionalism in the towers of the Brooklyn Bridge, wrote that the Williamsburg Bridge showed "a distinct artistic retrogression" from the earlier bridge. Noting that the chief engineer had minimized architectural detail, probably for economic reasons, Schuyler questioned why so much money is poured into palatial railroad stations when bridges are an even more prominent feature of the urban landscape. In an article in *Architectural Record,* October, 1905, Schuyler said "the designer has done his worst to make his bridge look ugly by the hugeness and insistence of the stiffening truss, which seems less like an ancillary construction than a rival construction. Outside of what had to be right in the Williamsburg Bridge [by which he meant the graceful line of the catenary cables], it may be said that everything is wrong."

Leffert L. Buck, original Chief Engineer of the Williamsburg Bridge, and, under Lindenthal, Consulting Engineer, was known for his courage and spirit of adventure. A native of upstate New York, he had spent 4½ years of his youth working in a machine shop. He had a reputation for dexterity, and was fascinated by the quick-paced changes in 19th century technology. When the Civil War started, Buck, a gaunt, rugged youth, joined the 60th New York Volunteers as a private. At war's end Buck had received several citations for exceptional bravery and had earned the rank of captain. After his successes in the Army, Buck decided to expand his mechanical knowhow. He attended Rensselaer

COURTESY OF THE NEW YORK TRANSPORTATION ADMINISTRATION

Above: Prodigious stiffening truss is emphasized. Note side spans are not suspended.

Left: Scientific American *drawing indicates site of new bridge (1896).*

Polytechnic Institute, graduating in 1868. When Washington Roebling asked that institute, his alma mater, to send some graduates to assist him in the construction of the Brooklyn Bridge, Buck was sent as one of the most promising. His connection with the Roebling bridges also includes improvements he was to make on John Roebling's suspension bridge across the Niagara River.

After the Brooklyn project, Buck went to South America to work on railroad bridges for the Oroya Railroad in Peru. His reputation for building sound structural designs over the most precarious passes in the Andes was well known in United States engineering circles. As a consultant, he made several trips back to the United States and did work for railroads in the Pacific Northwest, but he generally worked in Peru, Ecuador and Mexico.

Buck had neither the well-developed aesthetic sense of John Roebling and Gustav Lindenthal, nor, according to David Steinman, the patience for detail of Washington Roebling. Steinman notes that although Roebling had taken the precaution of galvanizing the cables of the Brooklyn Bridge to protect them against possible corosion, Buck omitted this detail in the Williamsburg Bridge.

However, Buck was confident of his engineering capability, and was easily able to call upon assistance from colleagues from Rensselaer. It is said that he often, to his credit, made use of their advice.

The new bridge certainly presented special problems. Traffic on the Brooklyn Bridge had increased beyond all expectations and ferry traffic had not diminished. Anticipating growing commuter traffic, the legislature required the new bridge to have a much greater capacity. They specified six lanes for elevated railroad tracks or trolleys, two for carriageways and a pedestrian lane. It was to be built on two levels and was to be half again as wide as the Brooklyn Bridge. The greater strength required presented a number of structural dilemmas. No radical change from the Brooklyn Bridge was felt to be warranted without good reason, yet no feature of the older structure was to be copied unless it was impossible to substantially improve upon it.

Buck's final plans, submitted in May 1897, provided for a span of 1600 feet between its towers—4.5 feet longer than the Brooklyn Bridge. The additional length is due to the fact that the land on either end is somewhat lower than at the Brooklyn Bridge. The bridge uses shorter cables as they do not have the function of holding up the roadways between the anchorages and the towers. But the cables, composed of 37 strands of 208 wires each, were much thicker than those on the Brooklyn Bridge. The two ends of the roadway are supported by steel viaducts. This was another cause for criticism since the roadway does not have a continuous curve from beginning to end, which is considered one of the most attractive features of suspension bridges. To carry all the tracks and roadways through masonry towers would have required towers of enormous width and weight, so steel towers were decided upon. These towers could be carried more easily to a greater height. The stiffening trusses, which range from anchorage to anchorage, ensure the bridge of an enormous strength, but pedestrians, crossing on the upper level tend to feel hemmed in by them.

These trusses are 40 feet high, probably a record, and one can imagine Buck following the advice of the famous English bridgebuilder, Thomas Smeaton, who once told an engineering student, "When you build a bridge, make it strong, sir, make it strong, and when you have got it strong, make it damn strong."

Unless Buck was influenced by the Eiffel Tower, he seems to have disregarded all architectural tradi-

PHOTO BY PETER ROSE

Upper level of Williamsburg Bridge features wide walkway. Note WB emblem on righthand side rail.

PHOTO BY PETER ROSE

tion. The few attempts at ornamentation on the Williamsburg Bridge are not integrated into the structure as a whole. At intervals along the balustrades of the upper level the letters W B are interlaced in an elaborate emblem. Where the towers rise above the second level of the bridge, there is a series of steel curlicues overlaid on the towers in an art-noveau fashion. The results are droll, as the decorations on the stolid bridge are so obviously an afterthought. They are actually the works of Henry Hornbostel, who also designed the Manhattan approach to the bridge and insisted on the widening of Delancey Street. Hornbostel, who also designed the balconies at each tower, wanted to top each tower with pinnacles which would contain masses of electric lights surmounted by large sculpted eagles. Hornbostel, who later complained of the difficulty of working within the Bridge Department, was overruled on the idea for the pinnacles. He had a freer hand when hired to do architectural details on the Queensboro and Hell Gate bridges.

At any rate, Greater New York, on December 19th, 1903, took the occasion of the opening of the bridge to congratulate itself on having consolidated and on having built the longest suspension bridge in the world. And the people of Williamsburg, on the Brooklyn side of the river, who were the prime movers in getting the bridge built, cheered their own great efforts. Williamsburg politicians had proposed the bridge in 1883, the same year that the Brooklyn Bridge opened, but they had been delayed for over a decade by ferry interests.

Williamsburg was named after Colonel Jonathan Williams, a U. S. engineer who was a grand-nephew of Benjamin Franklin. He surveyed the farmland between Wallabout Bay, where the Brooklyn Navy Yard was situated, and Newtown Creek, which separates Kings from Queens counties. The area was incorporated as a village in 1827, and soon became a bustling semi-industrial, semi-residential community surrounded by farmlands. It enjoyed its own commerce and ferry service with New York, and won incorporation as a city from the New York State Legislature for three years—from 1852 to 1855. After that it was annexed by its larger neighbor, Brooklyn. This consolidation made Brooklyn the third largest city in America.

However, Williamsburg still had its own impressive city hall, its own newspapers and history, its own banks. The largely German and Irish immigrant or first generation population was insular, and thought of their interests as separate from those of Brooklyn. In fact, they felt more allied to New York. They were geographically isolated from Brooklyn also, due to Wallabout Bay and its surrounding marshes.

PHOTO BY PETER ROSE

Henry Hornbostel was responsible for adding decorations of this kind to the Williamsburg Bridge. These are located beside the towers.

After annexation, Williamsburg residents noticed a diminishing rate of growth, compared with that of Brooklyn. Both in population size and economic expansion, Williamsburg was being overshadowed by its neighbor. The advent of the Brooklyn Bridge with the corresponding rising property values in that section of Brooklyn and the number of New Yorkers moving there, as well as the prospect of additional commerce in that area, aroused envy and anger among the citizens of Brooklyn's Eastern District, as Williamsburg was called.

On April 9, 1882, before the official opening of the Brooklyn Bridge, prominent residents of the Eastern District of Brooklyn held a meeting at the home of Captain Albert Snow to discuss the building of another bridge across the East River from Grand Street, New York, to Broadway in Williamsburg.

About 25 persons were presents, including ex-judges, an ex-sheriff and an ex-alderman. The consensus was that the Brooklyn Bridge, especially if no toll were imposed, would have the effect of benefiting one part of Brooklyn at the expense of the other. They also contended that making the Brooklyn Bridge free would add a burden to already overtaxed Brooklynites, and the benefits would only accrue to businesses and residents in the southern and central portions of Brooklyn.

An additional bridge, Williamsburg's leaders claimed, would complete and equalize the accommodations of all residents of Brooklyn who had occasion to frequently and regularly cross the East River. They urged Republicans and Democrats to unite against the election to the state legislature of any candidate who would not pledge to back the new bridge. The meeting also resolved to draft a plan for a permanent organization and call a mass meeting to publicize the need for the new bridge.

The meeting stirred up public opinion, but the situation remained dormant for a few years. Not until 1887 did it receive legislative consideration. At that time Thomas Farrell introduced a bill designed to promote the construction of a bridge or tunnel at approximately the present site of the Williamsburg Bridge. The bill was passed, but was not followed up by constructive measures or appropriations and, thus, brought no results.

In 1888, State Senator Patrick McCarren, a longtime Williamsburg resident, started a battle in the Albany legislature for the bridge. He found that he had not only the powerful ferry interests to fight, but also that New York was unwilling to assume one cent of the cost of the proposed bridge.

The first bill McCarren introduced in the legislature proposed that Brooklyn assume two-thirds of the cost of the bridge, while New York provide one-third. It was defeated. A spate of bills to snare a bridge between Williamsburg and New York followed, but all failed. The various measures were reported in the Williamsburg newspapers, and the local politicians assured the people that they were entitled to get the bridge. Eventually the population of Williamsburg was convinced that somehow, someday, their community would be inevitably graced by its own bridge. In the schools, in churches and at political rallies, it was repeated that Williamsburg could not prosper without its own bridge.

In 1892 it looked like Williamsburg would finally get its bridge. Frederick Uhlmann, owner of a number of Brooklyn's railways, decided his profits would improve considerably if he extended his trains over the East River into Manhattan, crossing the island to manufacturing districts on the West Side and sending routes down to the business district at Wall Street.

Hiring several well-known engineers, Uhlmann arrived at a plan which called for two bridges: one would cross the East River near where the Williamsburg Bridge now stands; the other would cross from a point near Gouverneur Slip, reaching Brooklyn south of the Navy Yard. Although the plan was at first scuttled in the legislature by the ferry interests, Senator Pat McCarren and Assemblyman Timothy D. Sullivan of Manhattan began a vigorous campaign to secure passage of a bridge bill. On March 9, 1892, the legislature approved a bill which founded the East River Bridge Company for the purpose of building Uhlmann's bridges. The bill provided that the Grand Street crossing had to be a suspension bridge and must carry, in addition to the desired elevated train tracks, roads for carriages and pedestrians.

The second bridge, which would have recrossed the East River at Hudson Street, Brooklyn, was permitted to be of cantilever construction, and would carry only trains.

However, the fact that Uhlmann intended to build his elevated railroad line across Manhattan infuriated the elevated interests in New York. They took Uhlmann to court, where the project was strangled in early 1894.

The battle which kept Uhlmann from building the bridges aroused renewed public sentiment in Williamsburg. New interests and politicians began to promote the idea. Guided by Tim Sullivan, State Senator George Owens and the executive committee of the People's Bank of Brooklyn, the People's Bridge Association was formed of prominent members of Williamsburg's Board of Trade and businessmen's organizations.

That spring a new plan was taken to the Albany Legislature, which, having passed two similar bills that had achieved no results, quickly approved the formation of a commission for the second East River Bridge.

The new bill, signed into law on May 27, 1895,

COURTESY OF THE COLLECTION OF PHILIP EAGAN
Williamsburg side of span seen before cables were strung.

provided for the members of the bridge company to purchase from any corporation that already possessed a valid charter, the right of that corporation to construct the proposed bridge. This meant that the New East River Bridge Commission would have to come to terms with Uhlmann, either getting the rights by mutual consent or by condemnation.

Uhlmann demanded a flat price of $650,000 for his company's franchise to build the bridge from Grand Street to Broadway. He was also willing to make a deal. The new bridge company could buy up his rights for $250,000 if they would agree to permit trains from his company—the Brooklyn Union Elevated Railroad—to cross the bridge free of all tolls in perpetuity.

The bridge company rejected both offers and threatened condemnation proceedings. Eventually, Uhlmann agreed to relinquish all rights for $200,000. Although this price was deemed unreasonable by the Bridge Commission, they agreed to it rather than have to take Uhlmann to court, which might delay the building of the bridge for years. This dispute with Uhlmann, however, may have contributed to the difficulty the city had in arranging for mass transit for the bridge. For several years after its opening, the Williamsburg Bridge could only be crossed by pedestrians, bicyclists and those with their own carriages.

Despite the early discouragements, the bridge was completed in only seven years—half the time it took for the construction of the Brooklyn Bridge. Its construction, which made use of all the technological advances of the preceding years, was marred by a severe fire on Nov. 10, 1902, which caused some doubts concerning the strength of the cables.

The fire started in a worker shack set atop one of the towers, and spread to the cables. Some wires were damaged. However, the Roebling Company, which had made the wire rope and woven the cables, had provided wire that was considerably above the strength required by Buck. The engineer decided to splice wires that had even greater strength at the burned out section.

The opening ceremony on Saturday, Dec. 19, 1903, marked by jollifications, decorations and parades on the Brooklyn side, was a municipal, not a national affair. Seth Low, who had been the borough president of Brooklyn when the Brooklyn Bridge was opened, was now the mayor of Greater New York. Mayor Low, accompanied by Manhattan Borough President Cantor and a delegation of commissioners, strode up the flag-draped bridge from the Manhattan entrance at Clinton and Delancey Streets. At the center of the bridge they met and shook hands with J. Edward Swanstrom, the Borough President of Brooklyn, Bridge Commissioner Lindenthal, Leffert L. Buck, and other influential citizens of Brooklyn. A regiment of 1400 policemen from both boroughs was on hand to prevent the kind of panic which had occurred after the opening of the Brooklyn Bridge. The congregation of notables turned toward the Brooklyn entrance and arranged themselves on a podium which had been set up at the Brooklyn plaza.

The passage of 20 years between the opening of the two bridges had seen much progress, both in the art of bridgebuilding and in the city politic. Gustav Lindenthal, who presented the bridge to the city, stressed these changes in his speech.

"The first bridge," Lindenthal said, turning to politics, "was built to connect two cities—New York and Brooklyn. This second structure over the East River is the first one built by the present consolidated great city, New York, which is destined to become the largest city in the history of mankind. It will have a distinction all its own, aside from its size. Our city will be preeminently the city of great bridges, representing emphatically for centuries to come the civilization of our age, the age of iron and steel."

Lindenthal was already hard at work designing the Manhattan and Queensboro bridges, which had been approved by the city in the past year. He took confidence from the fact that this new "colossal" structure had been completed after only seven years, and that it was reasonable to assume that the new bridges would take half the time it took to build this one.

Mayor Low, accepting the bridge on behalf of the city, also noted some alterations which had taken place during the intervening years. "The change in the conception of the function of an East River bridge and in political relationships of the communities which such a bridge is to serve are even more interesting than the changes in mechanical construction. When the Brooklyn Bridge was opened, two cities which before had been divided by a river, were for the first time connected by a bridge. This physical tie, slight as it was, proved to be prophetic of their manifest destiny. Today, the Williamsburg Bridge, although it was begun when the two cities were still politically distinct, unites not two cities but two boroughs of the same city. The East River has become a highway running through the city instead of a stream dividing two cities from each other."

Mayor Low, stressing the increased number of passengers flowing into Manhattan from Brooklyn daily, sadly noted the lack of mass transit facilities contracted for the bridge. He emphasized the fact that the new bridge would cater to through traffic, if possible, and not have the kind of terminals which had been provided the Brooklyn Bridge.

At 6:00 the next morning the bridge officially opened to the public at large. A crowd of bridge freaks and potential record holders gathered in the pitch black cold, vying to be the first to cross the bridge. A man named Wally Owen, who had been waiting since midnight, received the distinction. Owen raced across the bridge in a shiny new 56 horsepower auto. He made the round trip from Brooklyn to Manhattan and back in 6 minutes 50 seconds. The reporter who accompanied him on the trip said Owen nearly mowed down several police-

COURTESY OF THE NEW YORK PUBLIC LIBRARY

Falsework or scaffolding was used to erect side spans. They rest on steel arches instead of being suspended from cables.

COURTESY OF THE NEW YORK TRANSPORTATION ADMINISTRATION

Above: A severe fire in 1902 began in a workers' shack set atop one tower. It occurred while the cable was being spun. Right: The fire caused damage to both cable and footwalks. New stronger wires were spliced into the burned out section.

men. A few minutes later the first pedestrians began crossing from both ends of the bridge. Although the wind was howling and singing in the giant cables, there were wild footraces among the ambitious ones to be the first regular pedestrian across.

A New York Times reporter sent to watch the fray wrote that traffic crossing the Williamsburg compared favorably with the amount that regularly crossed the Brooklyn Bridge. The reporter noted,

> "There were all kinds of pioneers on bridge crossings on hand during the day. A bicyclist succeeded with much difficulty crossing the structure from end to end while riding his machine backwards. Another performed the feat of hopping across the structure, although he was compelled to halt frequently and get his wind ... The first one-legged man arrived at about 9:00 and stumped his way over for a record. At 6:20 the first dog trotted across the bridge from the Manhattan side ... No living person has ever yet succeeded in crossing the new structure on his or her hands. A bid for this distinction was made yesterday afternoon but the man only managed about 700 yards in an inverted posture."

By the end of the day, the first couple had kissed on the bridge during the freezing evening and one jaunty fellow brought a flask to become the first pedestrian to get drunk on the bridge.

Even after the first day of curiosity seekers, especially during rush hour, the upper level of the Williamsburg Bridge was consistently crowded. In fact, pedestrian traffic was so heavy that some politicians suggested New York's first moving platform be installed on the bridge. The crowding was due to the lack of any transit line across the bridge.

COURTESY OF THE NEW YORK TRANSPORTATION ADMINISTRATION

The city had great difficulty coming to terms with the privately owned elevated lines, subways and trolley routes in both boroughs. The establishment of an efficient connecting transportation system waited five years. In 1908 the Brooklyn Elevated System began running trains on the bridge's tracks.

Another factor considered unfortunate by some Williamsburg residents was the influx of Jewish immigrants into their predominantly German enclave, sometimes referred to as "Kleine Deutschland."

Before the bridge opened, a few orthodox Jews had settled in Williamsburg. They wore black coats, fur hats and wild beards and were looked upon as strange anarchists. Most of New York's Jewish refugee population was suffocating in squalid tenements on the East Side of Manhattan. When the bridge opened a considerable proportion rushed into the open space of Williamsburg and Brownsville. The Germans, in turn, moved down to Richmond Hill and Jamaica, leaving Williamsburg to the Jews and those too poor to move away.

Manhattan Borough President Jacob Cantor, speaking at the opening ceremonies, echoed this desire for the Jews to move into a more suburban Brooklyn setting. He considered it one of the prime benefits of the new bridge. "The erection of this bridge," he said, "has already had a beneficial effect upon the population of the east side of the Borough of Manhattan. It has, in fact, revolutionized the unhappy conditions existing near the bridge in that borough. It will ultimately, in my judgment, be the means of shifting the population from the congested districts of the East Side—or at least some of it—to more healthful and better homes elsewhere, with surroundings more agreeable, conditions more favorable and an environment more natural and wholesome."

Betty Smith's *A Tree Grows in Brooklyn,* published in 1943, portrays the attitude that the old Williamsburg residents had toward the growing Jewish community. The book unfolds the story of a half-Austrian, half-Irish girl named Francie, growing up before the first World War. It talks of the bridge, which she can see from the roof of her building, as a means of escaping the domestic, limiting world of Brooklyn. However, as a young lady crossing into Manhattan, for the first time, Francie is disillusioned with the bridge and the atmosphere of Manhattan. When Francie falls in love with a soldier during the war, they take a romantic stroll over the Brooklyn Bridge, not the Williamsburg.

Six months after the bridge opened, in July, 1904, the Woodbury Free Fruit and Vegetable Market opened under the landside arches of the Williamsburg Bridge. The market was part of a campaign to remove peddlers from the crowded dirty East Side streets. The market was between Ridge and Pitt Streets.

The neighborhoods at both ends of the bridge have declined as business centers since the beginning of the century. Now few pedestrians cross the Williamsburg Bridge. In 1964 The New York World Telegram and Sun reported that the bridge was so neglected that rust rained down on pedestrians, the guard rails were corroded and loose in places and benches were either broken or so flaked with rust that they could not be used. The only fresh paint, the paper noted, was in graffiti written by vandals. Although some steps have been taken to restore the bridge to its old condition, a complete refinishing job would be very expensive, and the pedestrian path is considered a hazardous zone even by the bridge's maintenance staff, one of whom was mugged on the path while doing his job.

When a tablet honoring the bridge's builders was

COURTESY OF THE ELECTRIC RAILROADERS' ASSOCIATION

Manhattan -bound train passes under plaque on upper level of Williamsburg Bridge. The name of Bridge Commissioner Gustav Lindenthal was omitted from this list of those who built the bridge. The bridge opened to the public in December 1903 but trains did not cross the span until 1908.

dedicated in 1904, Gustav Lindenthal's name was omitted. That year, the new bridge commissioner said in an interview that he did not know the reason. It may be that Lindenthal didn't want his name associated with the unattractive bridge. It also may reflect the difficulties he encountered with city officials during his tenure as bridge commissioner.

One cheery aspect of the Williamsburg Bridge is that it is the only bridge in New York designed with a special road for bicycles. The path is on the upper level. After all, the bridge was designed during the Gay Nineties, and, as one historian has said, "As much as anything else it was the bicycle that made the '90s gay."

Scientific American *featured this drawing on its cover in 1903. The cross section of the Manhattan approach shows an artist's conception of how all levels would look once transportation was in operation.*

COURTESY OF THE MUSEUM OF THE CITY OF NEW YORK

The steamer "Middletown" glides under the Manhattan span of the Queensboro Bridge in 1910, a year after the bridge opened.

QUEENSBORO BRIDGE AND MANHATTAN BRIDGE

THE BACKROOM BOYS BUILD BRIDGES

COURTESY OF THE BROOKLYN PUBLIC LIBRARY BROOKLYN COLLECTION

Manhattan Bridge towers take shape in 1903. For the first time towers had a 2 rather than a 3-dimensional configuration.

By the beginning of the 20th century, the science of long span bridgebuilding had gained acceptance from the public. Engineering journals and technological publications were heralding new discoveries in bridge techniques. Bridgebuilders were no longer considered impractical semi-magicians, but craftsmen and scientists, trained at approved technological institutions. However, in the planning of New York's next two bridges—the Queensboro and Manhattan—a conflict emerged as to whether bridges should be aesthetic as well as structural achievements. The chief proponent of introducing architectural detail into bridgebuilding was Gustav Lindenthal, who became the city bridge commissioner in 1902. Responsible for much that is handsome in the Queensboro and Manhattan spans, Lindenthal told a meeting of the Municipal Art Society that "In a bridge it is not possible to separate the architectural from the engineering features."

Of the two bridges, the Queensboro is more directly Lindenthal's design. He collaborated with architect Henry Hornbostel to design a cantilever span which could match the fantasies of Walt Disney. Despite its enormous size, its lines are gentle and graceful, and its fairytale spires make it an elegant portal to Manhattan's wealthy upper East Side. Its strong, immense spans, crossing the East River at Roosevelt Island, enhance the view from Sutton Place and United Nations Plaza. From the bridge's entrance in industrial Long Island City, midtown Manhattan at night glows like the Emerald City of Oz.

Further downtown, near Chinatown, the Manhattan Bridge also includes architectural embellishments. A Baroque arch and curved colonnade frame the entrance to the suspension bridge at the Bowery and Canal Street. They reflect Lindenthal's attitude, but do not bear his stamp. The arch was modeled after Porte St. Denis, a gateway to Paris, and the colonnade after the Bernini colonnade at St. Peter's in Rome. The architects were Carrere and Hastings, famous for the design of the New York Public Library at 42nd St. Carrere and Hastings were called in after Lindenthal and Hornbostel were dismissed in 1904. The formal Beaux Arts-style arch and pillars seem incongruous in today's bustling, grubby, downtown district. But they do provide a refreshingly odd backdrop for trailer trucks roaring from Brooklyn toward the Holland Tunnel, and an amusing counterpoint to the Italian Renaissance-style cast iron buildings of the Soho district. Isolated on an island facing Chinatown, the granite structures make one pause. Certain Department of Transportation personnel wanted to rip them down to make more leeway for the new Second Avenue Subway, but they are a part of the New York scene and will remain.

They are not the only artistically distinctive features of the bridge. The steelwork was also designed for aesthetic effect. However, Lindenthal's plans for the structure were scrapped by the city and Leon Moisieff (whose career was later ruined when the Tacoma Narrows Bridge fell) was chosen designer. Due to a lack of three-dimensional calculations by the designer, the Manhattan Bridge is the most troubled suspension bridge in the city and is now the subject of a controversy over how it should be strengthened. Workers on the bridge say the vibration is so great that mercury vapor lamps which last three months on the Brooklyn Bridge must be replaced every few weeks on the Manhattan Bridge.

Lindenthal, when he left office as bridge commissioner, warned that the new design would never stand up under the traffic for which it was intended.

Lindenthal was a crusading visionary and an excellent engineer. Appointed commissioner by reform mayor Seth Low, Lindenthal's eagerness for innovation and his convictions about aesthetics caused him to engage in battles with Tammany bigwigs and prominent citizens, among them the Roeblings.

While propounding the fusion of architecture and engineering, Lindenthal generally was careful of costs. Yet he had daring ideas: a 3100 foot suspension span across the Hudson which would carry 14 railroad tracks; the use of nickel steel eyebars instead of wires for the cables of the Manhattan Bridge and the use of the anchorages as revenue-producing auditoriums. History has proven him right as to the feasibility of these schemes. However, New York politicos did not like his insistence upon innovation, nor did the more conservative members of the bridge department. Although a committee of engineers found his plans sound, his designs were neatly sabotaged. After Mayor McClellan was elected in 1904, George Best was appointed to replace Lindenthal. Best managed to upset Lindenthal's plans for an eyebar suspension bridge and urged Moisieff to submit plans. Lindenthal was defeated, and a lengthy argument ensued on the merits of the bridge in newspapers and in professional journals. Most of the architectural features Lindenthal had proposed were finally incorporated into the bridge, in spirit if not in detail.

Lindenthal's career parallels that of John Roebling. Like his predecessor, Lindenthal received his engineering education in Europe. An Austrian, he attended the Polytechnic Institute, Vienna. His work in Europe included a stint as assistant engineer for the Austrian Empress Elizabeth Railway, and a job as bridge surveyor for the Swiss National Railways.

He arrived in America in 1874 at age 24. After serving as a consulting engineer for the Centennial Exposition at Philadelphia from 1874 to 1877, he found, as Roebling had, that Western Pennsylvania offered the greatest opportunities in bridge construction. He worked for the railroads of Pennsylvania until 1890, when he became a bridge engineer on a railroad which is now a part of the Erie-Lackawanna system.

Lindenthal came to New York in the late 1890's as engineer for the North River Bridge Co., which proposed to span the Hudson. He completed a design which would remain his unbuilt favorite for the rest of his life. Despite his later triumphant railroad bridge at Hell Gate, Lindenthal always remained unsatisfied because his North River Bridge was never built. Even after the completion of the George Washington Bridge, Lindenthal, at age 80, still advocated the building of his bridge at 57th Street. Had this enormous structure been built, it would even today be the suspension bridge with the greatest load capacity in the world. At the time he designed it, 3100 feet was an unprecedented length for a suspension bridge. It had the disadvantage of costing $130 million and it required deeper foundations than were then believed possible. Interestingly, its 600 foot towers were to be slightly altered versions of the Eiffel Tower.

Possibly, Lindenthal initially believed that once he was in office as New York City bridge commissioner he could convince the city that it should build his Hudson bridge, despite the refusal of the railroads. But his attention immediately was diverted to the Williamsburg Bridge, which was approaching an aesthetically unsatisfactory completion. And the Queensboro and Manhattan bridges were soon to occupy all of his efforts.

When Lindenthal was appointed, the latter two structures had already received financial appropriations and the Queensboro had been started. A design had been approved for the cantilever superstructure of the Queensboro Bridge. A suspension design had been specified for the Manhattan Bridge, but nothing definite had yet been accepted.

The Queensboro Bridge originally was known as the Blackwell's Island Bridge. Blackwell's (later Welfare, now Roosevelt) Island, had been rejected by John Roebling as a site for his suspension bridge in 1857. The island housed a city prison at the time. Ten years later, Long Island City's most influential citizens banded together to get permission to form a

COURTESY OF THE NEW YORK PUBLIC LIBRARY

Above: One of the earliest proposals for a Blackwell's Island Bridge was this iron-towered cantilever. Dr. Rainey suggested it stand at 77th St.

After three decades of politicking this conception of the New York and Long Island Bridge appeared on the cover of Scientific American.

bridge company. Among them were the Steinways, Pratts and Blisses, whose mansions lined the Queens waterfront. The New York State Legislature granted them the right to incorporate as the New York and Long Island Bridge Company for the purpose of bridging the East River at Blackwell's Island. The chief stockholder in the company was Austin Corbin, a railroad tycoon who had a majority interest in the Long Island Rail Road. He backed the bridge with the expectation that it would connect the Long Island Rail Road with the Harlem Line in Manhattan, obviating the need to transfer freight on ferries, a move which, incidentally, would make him a handsome profit.

The first plan for the bridge was submitted in 1868 by C. A. Trowbridge, a civil engineer who was president of the bridge company and vice president of the Novelty Iron Co. in Queens. He pictured two wrought iron cantilever spans over the river channels. The landside piers were to be built part way into the water so as to reduce the river spans to approximately 500 feet each. The connecting section of the bridge crossing Blackwell's Island would be a traditional trestle bridge. One roadway on the Long Island shore would lead to the Long Island Rail Road, and a second one would lead to the Brooklyn Navy Yard, in order to take advantage of profitable Brooklyn traffic.

Corbin met with some setbacks, got interested in other projects, and finally withdrew his support from the bridge company. The burden of reorganizing and refinancing the company fell on Dr. Thomas Rainey, treasurer of the company and one of its staunchest supporters. Dr. Rainey was confident the bridge would be a profitable investment. He visited financiers and politicians all over the East Coast to interest them in this investment. In a prospectus he prepared in 1874, Dr. Rainey noted that in addition to profits to be realized from commuters living in Queens and Northern Brooklyn, and the growing freight traffic from New England to these areas, Long Island was the most popular resort in America and would continue to grow in this regard with the aid of a bridge. He was also eager to cater to the funeral trade, as Long Island had 15 cemeteries at that time, and funeral carriages were a profitable business on the ferries. It was possible, he felt, that the bridge would enable Long Island to become an exclusive residential area. However, the fact that there was a relatively small population in the area probably mitigated against the successful completion of the bridge until Queens joined Greater New York in 1898. Even then Brooklyn residents who were experiencing a daily crush on the Brooklyn Bridge thought it extravagant to build a bridge to such a sparsely settled district. One Brooklyn Alderman charged that the bridge was being built "for prospective profits of certain landowners in the Borough of Queens." For all its wealth, the bridge company had succeeded in building one pier by 1893.

But in 1899 the Board of Alderman appropriated the necessary funds to really start work. The promise of better transportation had been one of the primary ways of getting citizens from the outlying regions to vote for affiliation with Greater New York. Now, fulfilling this pledge, the city fathers hoped to bind the city's union with steel wedding bands across the river. Appropriations were given helter-skelter for East River bridges 3 and 4 with little consideration to the practicality of plans. There may have been more interest in dishing out contracts to political favorites than in providing citizens wth transportation. Regarding East River Bridge number three (the Manhattan), the former mayor of Brooklyn, Schieren, pointed out that while he approved of building a second bridge to South Brooklyn, the designated site of the bridge was "from nowhere to nowhere," as there was no major road on the Brooklyn side.

In 1899 R. S. Buck submitted a design for the Blackwell's Island Bridge. This called for two cantilever spans of unequal length carrying two carriageways, two pedestrian paths, four trolley tracks and two railroad lines. It would be 120 feet wide. The cantilever construction was more economical than a suspension bridge as it made use of the intervening island. Altogether it took two years before the Army Corps of Engineers gave it approval. Meanwhile, a Committee of Forty representing Long Island business interests formed to expedite the building of the bridge. When the design received final approval from Washington in 1901 it was cause for jubilation.

When Lindenthal took over the post of bridge commissioner the following year, he was appalled by the abject ugliness of the design. He immediately hired Henry Hornbostel, then a prominent architect, to collaborate with him on remodeling it. Lindenthal intended to serve as his own chief engineer. Hornbostel's first duty had been to beautify the Williamsburg Bridge. The two then decided to redesign the Queensboro, retaining its cantilever construction but adding architectural refinement. Their first decision was to slim the bridge down to 80 feet by adding an extra deck. In the Lindenthal-Hornbostel design, the two uneven river spans took the shape of the universally admired curves of the Pont Mirabeau, over the Seine at Neuilly, except that the curve is over the roadway rather than under it.

The Lindenthal-Hornbostel changes, which included many architectural flourishes, aroused political opposition and much ill feeling. The Long Island Committee of Forty complained to the mayor about Lindenthal's arrogance and marched on the commissioner's office. Lindenthal managed to placate them, explaining the capacity of the bridge would remain the same as in the original design, pointing out his addition of elevators on both ends of the bridge would facilitate the transport of wagons. He also added elevators at Blackwell's Island.

One of Lindenthal's major accomplishments was to gain the endorsement of the Art Commission of the City of New York, which had up to then confined its activity to approving statues and municipal buildings. Joining Lindenthal's campaign to unite art and utility in bridge design, the Art Commission was instrumental in retaining much that is admirable in the Queensboro and Manhattan bridges despite the subsequent battles.

According to David Steinman, the eminent bridge builder, the design for the Queensboro Bridge is a clumsy, unfortunate imitation of a suspension bridge. Unlike most cantilever bridges, there is no suspended span between the two counterbalanced arms of the Queensboro spans. Thus it retains the lines of a suspension bridge. However, cantilever bridges had little tradition by way of design. Even today, of all the types of bridges, cantilevers take on the most diverse designs. In 1903, when Lindenthal designed the Queensboro, the most famous cantilever bridge was the Firth of Forth, a railroad bridge in Scotland. This was also the longest bridge in the world. While American architectural critics considered it an extraordinary feat of engineering, its appearance was considered too brutal for a city landscape. Thus Lindenthal's scheme of refinement was suited to the temper of his day.

Possibly when Steinman called the Queensboro clumsy, he was referring to the weighty tilted trusses which have the effect of an overgrown erector set. However, the responsibility for this must rest with

COURTESY OF THE NEW YORK PUBLIC LIBRARY

Architect Henry Hornbostel, who collaborated on bridge details, made this rendering.

COURTESY OF THE QUEENSBOROUGH PUBLIC LIBRARY

After Gustav Lindenthal become bridge commissioner he slimmed the roadway width down to 80 feet and added a second deck.

the United Pennsylvania Steel Co., contractors for the superstructure, rather than on the designer of the bridge.

An investigation of the stability and carrying capacity of the Queensboro Bridge in 1908 revealed that the bridge, which had been designed to carry four railroad tracks, four trolley tracks, two roadways and two walkways, would be overloaded from 25% to 40% if traffic on it reached its absolute maximum. Although these conditions were deemed highly unlikely, the Brooklyn Bridge had, in fact, met such enormous congestion and twice had suffered minor impairments. So the city was anxious to consider the possibility of maximum loading.

The main reason for this exhaustive investigation was the collapse of the Quebec Bridge, which shocked the engineering world. Started at the same time as the Queensboro Bridge, the Quebec Bridge was to have been the longest cantilever bridge in the Western Hemisphere. The bridge collapsed under its own weight, while under construction, killing 74 construction workers, mostly Mohawk Indians. David Steinman calls it the most startling catastrophe in bridge engineering history. An investigation indicated that the Phoenix Bridge Co. had tried to cut corners and had avoided preliminary testing. Thus, the designer had used methods perfectly acceptable in shorter cantilever spans but inadequate for such a long one. The fall and subsequent rebuilding of the Quebec Bridge remains one of the great dramas in engineering history. It had the unlooked-for benefit of promoting the scientific analysis of bridge design. Even now, members of Canada's Iron Ring Society wear rings which are forged from the metal of the collapsed bridge. They are worn as a reminder of the engineer's responsibility.

With the collapse of the Quebec Bridge, the Queensboro Bridge became temporarily the bridge with the longest cantilever span in this hemisphere. Although the $20 million New York was spending on the bridge was not indicative of any frugality, and although the longest span was only 2/3 the length of the Canadian structure, the collapse of the Quebec Bridge made New Yorkers suspicious. Led by the *New York Herald Tribune,* they clamored for an investigation. Eventually, two independent engineering consultants were hired to examine all facets of the bridge's construction. One was a private firm, the other a Columbia University engineering professor. Both reported that all the workmanship done was first rate, but both found that the design would place too much stress on certain members if loaded to peak capacity. They also argued that the design did not take into account

PHOTO BY PETER ROSE

extra weight from snow falls. The solution was simply to reduce the number of elevated tracks to two, and to remove some deadload, or unnecessary structural material. A rumor circulated among engineers which accused the United Pennsylvania Steel Company of purposely adding extra material to the bridge, as they were being paid by the pound, and this jacked up their profit. Perhaps the unauthorized deadload is what caused Henry Hornbostel to exclaim when he first saw the completed superstructure "My God—it's a blacksmith's shop." He couldn't believe it was his design.

The furor over the safety of the bridge raged on for months after the findings had been made. The *Times* defended the bridge; the *Tribune* denounced it. *Engineering News* agreed with the *Tribune; Engineering Record* backed the *Times. Scientific American* called the design the worst blunder in the history of bridge design. The arguments put forth were sometimes completely unscientific. One newspaper, for instance, ran a rather amusing interview with Edward E. Sinclair, an alleged engineer on the bridge. Under a headline reading BIRDS ON BRIDGE VOUCH ITS STRENGTH! WOULDN'T IF IT WERE WEAK, Mr. Sinclair was quoted as saying: "I have been bridge building for twenty years, but never before saw so many birds and so many kinds of them together on a bridge as are now at night on the Queensboro Bridge. Kipling in his *Bridge Builders* points out that whenever numbers of birds gather on a bridge in process of building or nearly completed, it is an indication that the structure will stand all tests. This has also been my experience... Ornithologists say that birds in large flocks will not settle on a weak structure."

The construction of the bridge was marked by use of a steel falsework, a novelty in those days, to build up the shore arms, which counterbalance the channel spans. The channel spans were cantilevered out with the aid of two traveling cranes. About 50,000 tons of steel were used, and it marked the first time that nickel steel, stronger but lighter than carbon steel, was used for tension members and pins.

An idea of the weight of the members may be gotten from a description of the methods of construction. "All material was floated to Blackwell's Island and unloaded with special 65-ton derricks, delivered to storage yards served by electric gantries having 65-ton main hoists and auxiliary hoists of 35 and 10 tons. Eye bars were then packed together and lifted as one member. Pins weighing 7,000 pounds apiece were driven with a two-ton ram. The rivets at chord bottom are one-inch in diameter with a seven-inch grip and the structural steel in the falsework and travelers added up to 4,000 tons. The heaviest bottom chord section weighs 90 tons and is 100 feet long."

The dead weight was soon removed and the bridge was opened for traffic in April 1909, although the official opening ceremonies were postponed until July 12.

By September, trolley cars of the Queensborough Bridge Railway began a short and simple shuttle service, going back and forth between stations at each end of the bridge. They used the outer lanes of the lower roadway. En route the cars stopped at two other stations: one above Vernon Boulevard on the edge of the river in Queens, the other above Welfare Island. From these stations passengers descended to street level in elevators. Other trolley lines shared use of the tracks at various times. Most terminated at an underground plaza in Manhattan, which still remains. In 1917 an elevated line to Corona and Astoria began running on the upper level of the bridge. This line soon proved inadequate for the growing Queens population and a subway had to be dug under the river at 60th Street. The elevated line was removed in 1942. The original trolley line lasted until 1955, when the last car ran over the bridge. That year the Welfare Island Bridge opened, providing a direct automobile route to the island and making the trolleys unnecessary. After that the bridge was converted exclusively to automobile traffic.

Utilitarian considerations aside, the "style" of the bridge was commended, with architect Hornbostel himself in the front ranks. In 1909 an article by Hornbostel appeared in the pages of *Architecture Magazine,* saying "It was no easy matter to break into the highly useful and practical atmosphere established by the engineer" and said many "modifications and unhappy alterations" had been made on his design. Nonetheless, he was pleased that many of his ideas were carried out. Admitting a measure of vanity, Hornbostel called attention to the ample roadways, "the unimpeded footwalk passing under a perspective of elliptical braces most cheerful; the four massive and gaily capped steel towers are very impressive; the rails and portals interesting... The cantilever has an interesting, attractive, graceful design." Hornbostel went on to list the points of architectural interest, starting from the New York approach: two bronze lanterns 18 feet high, probably the highest in New York; the little trolley entrances in cast iron and terra cotta; high portals with bronze tablets flanked by the granite staircase.

Another article in *Architecture and Builder* congratulated Hornbostel on his innovative design. "The structure of the bridge itself is an intricate mass of interlacing steelwork, seemingly incapable of architectural beauty because of the strict requirement imposed by the structural conditions in the design of compression and tension members. Yet, as we look upon the bridge from varying points of view, there is a charm, a certain gracefulness in the repetition of symmetrical parts, in the tilting struts which

COURTESY OF THE MUSEUM OF THE CITY OF NEW YO

A steel falsework was used to build the shore arms, while the overwater span was cantilevered out with a traveling crane. This photo shows work in progress in 1907. Opposite page: steel eyebars and pins. Span's structural steel weighs 7,000 tons.

spring from the sturdy granite towers and are lightly topped with finial ornaments. Wrought entirely of structural steel, at many points small adornments add appreciably to the delicacy of the structure."

Hornbostel's iron work decorations have been widely admired. Never before had iron work been treated on such a grand scale. Hornbostel was called "daring and inventive, never hesitating about executing work because it has no precedent."

The bridge, then, was accepted as a work of art, as well as a tool of travel. It makes it somewhat appropriate that it is the bridge most traveled upon by North Shore matrons in their excursions to the Guggenheim, Whitney or Plaza. F. Scott Fitzgerald in "The Great Gatsby" mentions the bridge's role in the travels of his elegant Long Islanders.

Exploring the Queensboro Bridge by foot was a delight until the recent closing of most of its walkway. Not only are there many unexpected decorations, but beneath the landside arches on the Manhattan shore there is a market place lined with Guastavino tile. The marketplace may soon be the setting of a cinematheque designed by I. M. Pei, the architect responsible for the World Trade Center.

Midway along the span the walkway made a descent to a catwalk which crossed the underbelly of the roadway and terminated on Welfare Island at the "upside down building," whose entrance lobby was on its eighth floor. For decades the elevators in this building were only means of access for vehicles to the island until a bridge was built in 1955 connecting it to Queens. The building, which was mainly entered through its roof, was featured in Ripley's *Believe It or Not.* However, the building has been demolished now that Roosevelt Island has been converted to a housing development. Transportation from the island to the Manhattan shore now takes the form of a cable car situated just north of the Queensboro Bridge.

The opening ceremonies of the Queensboro Bridge in 1909, sponsored by the 37 living members of the Committee of Forty, were an attempt to outdo the celebrations for the Williamsburg and Brooklyn Bridge. Manhattanites ho-hummed it, but after speeches by New York Governor Hughes and Dr. Rainey, the bridge was enveloped in fireworks for two hours, in a display which astounded even members of the press. One particular crowd pleaser, a

PHOTO BY PETER ROSE

New York Times reporter wrote, was an immense representation of Niagara Falls pictured in all the colors of the rainbow pouring over the bridge on the Queens side. Two tons of powder were said to have been used for this effect alone. Bloomingdale's held a sale in honor of the bridge. Representatives from several marathon runners' clubs raced along the roadway. A choir of young ladies from Queens presented a comic opera called the Mockingbird.

Today the bridge is still celebrated and worthy of admiration. Tracks no longer cross the bridge, but the little entrances to the underground trolley plazas remain. When Second Avenue was widened after the upper roadwy was converted to auto traffic, a plaza was to have been built by Alexander Delacorte, who donated the Alice in Wonderland statue and the Delacorte Theater in Central Park to the city. However, alternate plans by Robert Moses for parking facilities and office complexes, although they were never consummated, checked the project, probably to the detriment of that area; the wide cobble-stone exit on Second Avenue is an eyesore, compared with the plaza and balconies which lead to Delancey Street on the Manhattan side of the Williamsburg Bridge. Perhaps it is this wide cobblestoned street that Simon and Garfunkel mention in their *Fifty-Ninth Street Bridge* song.

Lin Yutang, the Chinese-American author, expressed his feelings for the Queensboro Bridge with eloquence in his *Chinatown Family,* a saga of Chinese immigrants. He wrote that his hero, Tom,

COURTESY OF THE NEW YORK TRANSPORTATION ADMINISTRATION

Left: These vaulted arches lined with Guastavino tiles are beneath the Manhattan shore span on the Queensboro.

Below: Architect I.M. Pei designed a cinematheque to be housed in those arches.

Opposite page: A huge crowd attended the opening of the new bridge to Long Island. Lavish entertainment included fireworks simulating Niagara Falls.

COURTESY OF THE NEW YORK TRANSPORTATION ADMINISTRATION

COURTESY OF THE AUGUST ROCHLITZ COLLECTION

"sometimes walked alone to the head of the Queensboro Bridge, drawn by a mysterious power like his early fascination with the El. The bridge leaped up some 60 feet above him, a mass of black crossed steel trusses, supported by heavy black stone towers that might vie with kings' tombs and medieval chateaux in their height and size. He saw that the bridge had the sweep of the sea and the grace and strength of a great work of art, and the independence and pride of a beautiful woman, the leaping power of a leopard."

One major fault in the span is that the five central lanes on the lower roadway make for more than the usual number of accidents. In fact, there were more accidents on this bridge than on any other in New York. Recently the Department of Transportation welded steel studs every few inches along the steel on the roadway; these studs reduce the slipperiness of the deck on rainy days. Another change made about two decades ago reduced the height of the bridge's spires. The hollow steelwork was so rusted inside, that the Bridge Department believed them dangerous.

While the Queensboro Bridge was under construction, a second more acrimonious controversy arose over the design of the Manhattan Bridge. Designated as a suspension bridge, the Manhattan was to have a span of 1470 feet between towers, somewhat less than the Brooklyn Bridge, a quarter of a mile to its south. However, including its shore spans it was to be the longest in the world. A plan presented by the Board of Engineers in 1901 called for the bridge to have four towers composed of heavy trusses and crowned by minaret-like ornaments. Along the roadways was to run a heavy stiffening truss 55 feet high, even higher than the universally condemned truss on the Williamsburg Bridge. The design never received official approval.

Lindenthal submitted a design in 1903 which did away with the uglier features of the traditional con-

COURTESY OF THE BROOKLYN PUBLIC LIBRARY BROOKLYN COLLECTION

The Second Avenue Elevated line ran on the upper level of the Queensboro Bridge. The photo was taken in 1931

cept. Besides rather elaborate ornamentation, the Lindenthal-Hornbostel design departed from conventional bridge building by introducing to America the concept of eye bar suspension bridges. The Elizabeth Bridge just being completed over the Danube in Budapest had shown this type of bridge to be a successful adaptation of suspension bridge design. The cables, instead of being woven from steel wires, would consist of four chains of nickel-steel eyebars, which would be stiffened and braced. From the trussed chains the roadway would be suspended. This gives the traveler an unimpeded view, as there are no stiffening trusses along the roadway. Construction by method of eye-bar chains also saves time, Lindenthal noted, since the bridge can be constructed by means of a falsework, instead of requiring the months and sometimes years it takes to weave the wire cables.

Lindenthal's design also utilized the space in the two anchorages. Each was to house great public meeting halls to be devoted to any public purpose the city might determine.

The towers, too, were a novelty. There were to be four columns in each tower, which would be lined up in a plane rather than placed in a three-dimensional configuration like the towers on the Williamsburg Bridge. This would allow for expansion, contraction and flexibility. For instead of the cables riding over on rigid towers, as in the other two sus-

COURTESY OF THE ELECTRIC RAILROADERS' ASSOCIATIO

Trolley from Manhattan at Welfare Island stop. Riders took an elevato from the island to the bridge. In 1954 a lift bridge from Welfare Is land to Queens was built and trolley service ended.

Scientific American *published this drawing of "Suspension Bridge Number 3" in 1901. Like the Brooklyn Bridge to the left, the new span would sport diagonal stays. The stiffening truss was enormous. The design was modified many times before construction.*

pension bridges, the towers of the new bridge would bend with the movement of the cables by means of pivots at their bottom. This was found to be advantageous because the rollers over which the cables moved in the Brooklyn and Williamsburg towers tended to rust and be immobilized by ice in winter. Each tower's four columns were to be held together by steel frames so that the entire fabric of the tower would have ample strength.

Where the towers met the roadways, there were to be balconies for recreational use in summer. Each tower was to have an ornamental cornice and decorative pinnacles.

The innovative engineering techniques suggested for the new bridge were met with some skepticism from politicians. Mayor Low appointed an impartial board consisting of five engineers to examine the soundness of the scheme. Although the engineers unanimously endorsed the design, and the Art Commission also approved it, the Board of Aldermen still withheld appropriations. Both the *New York Times* and *Scientific American* attributed this delay to Tammany politicians who wanted to secure contracts for the Roebling wire works. *Scientific American,* which heartily backed Lindenthal's eyebar design, said that the Roeblings had definitely been instrumental in delaying appropriation of funds for the new bridge. These tactics continued until the Tammany-backed McClellan was elected Mayor in 1904. His new bridge commissioner, George Best, stepped in to foil Lindenthal's design. Best appropriated some bridge commission money for a third design. This one was by Leon Moisieff.

The new design, which took about six months to prepare, reverted to the old wire cable method. It might be noted that there were eleven companies in the United States which could supply the nickel steel eyebars for Lindenthal's design, while only the Roeblings could be awarded the contract for the wire cable design. The new design exhibited some of the refinements of the Lindenthal plan in its well-formed steel towers and classical architecture. However, it also retained the much criticized stiffening truss, although this one, at 26 feet, was not as high as the one that had been proposed at first.

Lindenthal, now out of office, defended his plan in letters to the public in *The New York Times* and to his colleagues in professional journals. In advocating his own design, Lindenthal stated, "It was my intention to send complete plans for the Manhattan Bridge as well as the Queensboro Bridge to the Inter-

COURTESY OF THE MUSEUM OF THE CITY OF NEW YORK

The Manhattan Bridge with cables spun and before roadway was suspended. The photo was taken in 1907.

national Exhibition in St. Louis. I had every confidence that they would receive distinction for both design and making an advance in bridge engineering." He claimed that the new design, which was to carry the heaviest load ever conceived for a suspension bridge—two roadways, four trolley tracks and four elevated lines—needed the objectionable deep stiffening truss if it were to use the wire cable suspension method. Otherwise it would be too weak.

The public by this time was no doubt accustomed to squabbles among its engineers. But the hysterical denunciations which accompanied both sides of this argument led the *New York Times* in an editorial to quote Shakespeare's Hamlet,

> Tis the sport to see the engineer
> Hoist with his own petard.

They compared the energetic animosity of Lindenthal and Best to that of admirals in the Russo-Japanese War.

The *Times* sided with Lindenthal, defending his design on the grounds of its architectural qualities and the fact that it had already been approved by unbiased engineers. The newspaper also questioned the cost and delay of preparing a third plan. Meanwhile, the Lindenthal design was labeled by opponents as an architectural freak; they contended that it would also be unnecessarily expensive. This was never ascertained, for McClellan and Best, without providing an adequately detailed report, rushed Moisieff's design through the Art Commission, which approved the plan over the dissenting voice of its president. This time the Board of Aldermen had absolutely no qualms about appropriating almost twice the money that Lindenthal had requested. By the beginning of the next year, the first contract was awarded to a Tammany regular. The Municipal Art Commission later stated that it had inadequate guidelines for accepting bridge designs, as it would seem they must consider engineering and economic factors as well as the aesthetic ones, and were not well enough informed to make a total criticism of either design. They decided that as long as the new design adhered to the architectural effects Lindenthal had promoted, it was acceptable.

Although the eye bar cables were scrapped, the new design incorporated several notable Lindenthal-Hornbostel architectural features. The towers are flexible and lined up in a plane, though without the pivots. The anchorages, surmounted by a 40 foot granite colonnade, are replete with arches, buttresses, pilasters, cornices and balustrades. Through each anchorage passes a street—Water Street in Brooklyn and Cherry Street in Manhattan. The designer was also careful to ensure that at no point between anchorages does the graceful curve of the cables intersect the line of the stiffening truss.

A street runs through each anchorage of the Manhattan Bridge.

The 26-foot high truss and other aspects of the Manhattan Bridge have been scrutinized by David Steinman and other engineers, and have been found wanting. They echo the predictions of Lindenthal: the bridge cannot hold the traffic for which it was designed. Steinman in the 1940s pointedly suggested that the best way to save the bridge was to remove the subways and build a separate tunnel for them.

Since the bridge was finished several modifications have had to be made to compensate for the weakness in the design. "Moisieff was a brilliant man but he didn't have a computer to help him," according to Harold Samuelson, an engineer who has proposed a number of modifications to stiffen the Manhattan Bridge and make it suitable to carry subways on its outermost lane. At present, the trains cause a great strain on the outer cables, according to Samuelson, since the outer cables contribute the major source of torsion resistance. Maximum torsion occurs when subway trains start to cross the opposite sides of the bridge at the same time. Then one side of the roadway dips three to four feet while the other side rises a corresponding amount, making the relative deflection six or eight feet.

Although the average motorist may not feel this movement, it has caused strain on many parts of the bridge and has cost the city a large sum for renovations. The stringers of the upper roadways had to be replaced, and the bracing frequently failed despite numerous additions. It was finally removed altogether, when engineers decided that the bridge needed more room to breathe. There was also slip on the cable bands which attach the suspenders to the cables, and there was chafing of the suspenders where they passed through the upper chords of the stiffening truss to meet the roadway. Wooden buffers were installed to alleviate this.

Moisieff's design for the Manhattan Bridge was the first to make use of deflection theory, which results in economy of material. The theory allows that suspension bridges are stronger than they at first were considered, because the curve in the cables makes them more efficient in carrying loads than stiffer types of bridges. Thus the structure could be lightened. However the designer did not take into account the problem of the subway trains moving in the outer lanes. The Williamsburg Bridge was also designed to carry subway trains, but there they are on the inside lanes and the stiffening truss is much deeper.

Two recent studies of the bridge have presented methods to correct the distortion. The city commissioned the late David Steinman's firm of Steinman, Boynton, Gronquist and London to recommend improvements. With the cooperation of Columbia University's engineering school, the consultants constructed an elastic, two dimensional model, 1/50th the size of the real bridge. After experimentation with this model they concluded that extra columns below the roadway and some radiating stays, like those on the Brooklyn Bridge, would stiffen the bridge adequately.

The other study, privately made by Harold Samuelson for Computer Data Corporation, states that

COURTESY OF THE AUGUST ROCHLITZ COLLECTION

Above, aerial view of Manhattan Bridge taken in 1965 from Brooklyn side. Top level carries four auto lanes. Bottom level has tracks for subways and auto roadways. Below, Columbia University Engineering School constructed this two-dimensional model of the Manhattan Bridge to check the twist caused by subways in outer lanes.

COURTESY OF THE NEW YORK TRANSPORTATION ADMINISTRATION

COURTESY OF THE NEW YORK TRANSPORTATION ADMINISTRATION

The Manhattan side features an arch modeled after Porte St. Denis, an entrance to Paris.

the bridge should be stabilized by erecting eight torsion diaphragm posts, or trussed posts, which would extend from the inner lower level roadway to the cables and take advantage of the horizontal strength of the bridge by coupling it to the vertical strength. In addition, this study suggested stiffening trusses between the inner and outer cables, much as Lindenthal had suggested between his eye bar cables.

However, in 1909, in the bliss of ignorance, Mayor McClellan rushed the work, eager to prove to New Yorkers that a man allied with Tammany could build public works faster and cheaper than a reformer. The cables were spun in a record four months time. McClellan was present the day the weaving started and the day it ended, receiving congratulations and urging the workers on. Lindenthal, who was now working for the Pennsylvania Railroad and the New York Connecting Railroad, published a paper claiming that other bridges built under Tammany administrations took so long to build and went so far above their budgets because, before each election, the politicians would send to the construction site large groups of inexperienced men who would be authorized for jobs in order to curry votes. The engineers would then have to pay to keep them out of the way or else they slowed down the work.

McClellan didn't want that kind of publicity for the Manhattan Bridge. He insisted that it be finished before he left office on December 31, 1910. So, on that date, although the roadway was not paved and planks had to be laid down, a small quiet procession of automobiles crossed a partially finished bridge, declaring it open. It was McClellan's last act in office. It seemed a good omen that the span's first traffic consisted of holiday crowds bringing New Year gifts to Brooklyn friends.

Amtrak train about to pass through arched tower of Hell Gate Bridge. PHOTO BY VICTOR HAND

HELL GATE ARCH

LINDENTHAL VINDICATED

Gustav Lindenthal's mammoth Hell Gate Arch stands like a memorial to the days when the railway systems were the lifelines of the country. In its heydey, more than 60 trains crossed each way daily. Now over 30 Amtrak passenger trains and two freights cross the structure each day. This is an improvement. After the 1970 failure of the Penn-Central Railroad, motorists on the adjacent Triborough Bridge rarely saw trains pass over it at all. At that time only four passenger trains crossed it daily in each direction.

This underutilized bridge was completed in 1916. It was the longest steel arch in the world, surpassing its nearest rival by 200 feet. It is still the strongest long span bridge in the world. In fact, according to designer Lindenthal, the four-track bridge is strong enough to withstand the weight of 60 200-ton locomotives laid end to end. However, the very day the initial run was made on the Hell Gate Arch Bridge, an editorial in the *New York Times* foresaw the doom of the railroad monopoly on transportation and the belated character of this impressive bridge. It opened two days before Congress declared war on Germany in April 1917. The next year the Federal Government nationalized the operations of the railroads to facilitate the war effort. Although the railroads went back to corporate managements after the war, the railroad system in America has never quite been the same, probably due to the subsequent increased popularity of the automobile.

It is especially sad that this imposing and beautiful arch is almost a vestigial structure. Its prime authors, Alexander J. Cassatt, then president of the Pennsylvania Railroad, and designer Gustav Lindenthal, were eminently practical men, not apt to dream up projects for personal prestige.

Cassatt was president of the Pennsy from 1899 to 1907. When he took office, the Pennsylvania Railroad ended at a passenger terminal in Jersey City and a freight terminal in Greenville, New Jersey. Passengers and freight bound for New York City had to be ferried across the Hudson at considerable delay, as the river was one of the busiest waterways in the world. Cassatt devised the scheme of tunneling under the Hudson and bringing passenger trains into midtown Manhattan at what was to become Pennsylvania Station.

Cassatt called in Lindenthal when he decided to bring his trains across the Hudson, as he was uncertain at first whether to tunnel under the river or build a bridge. As early as 1886 Lindenthal had designed a six track suspension span for the Hudson to be placed at Desbrosses Street, but this was vetoed by the Army Corps of Engineers on the grounds that its piers would obstruct navigation. In 1902, Lindenthal submitted an improved plan which met the approval of the Army Engineers. Although the capacity of the bridge would have been greater than that of several tunnels, Cassatt was furious that other railroad companies would not contribute what he felt was a reasonable share of the cost of the bridge and were leaving a disproportionate burden of its financing to the Pennsy. The estimated cost of the project was $100 million. He decided to build a tunnel to which only Pennsylvania Railroad trains would have access. The twin tunnels to Penn Station, used exclusively by Penn. trains, were finished in 1910. So was the station.

Penn Station, one of the biggest and grandest in America's era of opulent railroad stations, was designed by McKim, Mead and White after the ancient Roman Baths of Caracalla and the Basilica of Constantine. The Roman Doric station with its 300 by 110 ft. waiting room was ripped down in 1960 to make room for the present Penn complex.

Although Lindenthal left the Pennsy temporarily in 1903 he was obsessed with the Hudson River Bridge project. He formed the North River Bridge Co., a corporation which he hoped would build a bridge over the Hudson, but he was never able to secure the necessary funding.

During Cassatt's presidency the Pennsylvania Railroad acquired a majority interest in many competing and minor railroads. Besides knitting together a vast web of former competitors in the Midwest, Cassatt also bought controlling interest in the Long Island Rail Road after the death of its major stockholder, Austin Corbin. Corbin had devised a number of schemes for extending the Long Island Rail Road into New York and up to New England. One was the Blackwell's Island Bridge, which was to have brought the Long Island trains into Manhattan to join up with the tracks of the New York Central on Park Avenue. Another was a tunnel under the East River. The most ambitious scheme, however, was the New York Connecting Railroad, which by means of extensive viaducts and bridges would link the Sunnyside Yards in Long Island City with railroads in Westchester and on up to New England. Along with the Long Island line, Cassatt's purchase included the rights to the proposed New York Connecting Railroad. The plan caught Cassatt's imagination and he suggested that the Pennsylvania Railroad build the viaduct system across Hell Gate between Queens and The Bronx.

Cassatt immediately ordered the digging of a tunnel under Manhattan and the East River which would bring the Long Island Rail Road into Penn Station. In anticipation of the building of the New York Connecting Railroad, Cassatt expanded some recently acquired freight yards and dock facilities at Bay Ridge in Brooklyn and made the Sunnyside Yards in Queens the largest passenger yards in the world.

These improvements and acquisitions brought about easier transfer from Long Island and New York City to the Pennsylvania's western and southern routes, but New England, home of much manufacturing, was still virtually isolated. Goods from New England reached The Bronx by way of the New Haven Railroad. From there cars had to go by ferry down the busy East River and across the Narrows to railroad yards in New Jersey. This 14-mile trip could take as long as three hours and there was always the hazard of a ferry capsizing. In severe winters ice often blocked the waterways.

Passengers from New England destined for points south of New York City, along what is now called the Northeast Corridor, also had their problems despite the construction of Penn Station. After boarding the New Haven Railroad at a New England city, the passenger would arrive at Grand Central Terminal, hastily collect his baggage and race across town in a taxi to Penn Station to catch the southbound train. There was an alternative to the New York City run. This route crossed the Poughkeepsie Bridge, but most passengers are reported to have avoided the northern route as this was several hours slower.

Although the New York Connecting Railroad was promoted by Cassatt as a way to obtain direct freight and passenger service between New England and the rest of the country, it was not built until 1913, when Samuel Rea became president of the Pennsylvania Railroad. Rea, who had an engineering background himself, had played a part in the building of the Pennsylvania's tunnels under the Hudson River. Cassatt had sent Rea to England to investigate methods of tunneling being developed there. During the controversy about whether the Hudson should be crossed by bridge or tunnel, Rea had met Gustav Lindenthal, who at the time had not yet come under the strain of being New York City Bridge Commissioner. Rea had been impressed with the man. More than 10 years had passed since the controversy over the tunnels, and although Lindenthal was in bad favor with New York's politicians, the new president of the Pennsylvania Railroad was not one to respect their values and judgments.

He hired Lindenthal for the new project and gave him complete control. The new connecting link was to leave the New Haven's Harlem River branch at 142nd Street in The Bronx and, in a sweeping curve, cross the Bronx Kill to Randalls Island, then the Little Hell Gate to Wards Island, then running along the easterly edge of Wards Island, cross the Hell Gate itself. On the Queens side it was, of course, to connect with the Sunnyside Yards, and from there passenger trains would run under the East River to Penn Station. Freight trains were to continue through Queens on to the old Manhattan Beach line in Brooklyn, which ran to the Bay Ridge docks. After that it was only a short ferry trip over the narrows to New Jersey. Rea envisioned an underwater tunnel here too, which would have provided a direct freight run to Greenville.

The Hell Gate Bridge was the culmination of Gustav Lindenthal's career. It is the only example in New York of a span both designed and supervised by this master bridgebuilder. One can't help admiring every detail. It is graceful despite its great weight. The difficult conditions of the waterway required meticulous engineering. Lindenthal may even have welcomed these difficulties as a great challenge, for that was his nature.

He believed that an engineer, given enough money, could do practically anything. In an interview in *Frank Leslie's Magazine* in 1920, Lindenthal

COURTESY OF THE MUSEUM OF THE CITY OF NEW YORK

Artist Richard Rummell did this rendering of the Hell Gate Arch in 1907, 7 years before construction began. At the time plans showed the arch abutments separated above the roadway. View shows Queensboro Bridge in the background.

boasted he "could build a bridge across the Atlantic and have piers on a solid foundation even though in places the ocean is three miles deep." Such a bridge would be built 300 feet high on floating, anchored islands and "would be strong enough to carry the heaviest traffic and to resist the biggest gales that have ever blown. There is nothing at all impossible in such a project."

An anecdote from the bridgebuilder's career will illustrate his belief in his own ability and his doggedness in convincing others to trust his judgment. Lindenthal had come to Amerca in 1874, not from any political necessity, as had John Roebling, but because he was intrigued by the experimentation and advances in bridgebuilding in America. His English was not good, but he toured several of the major engineering works of America and wound up back in New York with only $150 left. He had not only had an academic training in engineering, but also had worked for his father during summers as a mason and bricklayer. As the Philadelphia Centennial Exposition needed manual workers that year, Lindenthal signed on as a mason on one of the buildings. However, instead of living like a workman, he took a room at an expensive hotel, dressed well and took English lessons every evening. He let word get around that he was a capable draftsman, and when a draftsman left the office of one architectural firm, Lindenthal got the job. In three days he completed an assignment expected to take three weeks. He was hired permanently. When the firm's chief architect spoke with the young Austrian he realized Lindenthal was an inspired and well-trained engineer. They asked him to submit a design for the exposition's principal building. Lindenthal's design outstripped all the competition and was used for the building.

"Failure is only an epitaph for lack of preparation," Lindenthal was to say later. His belief in the engineering sciences as capable of all undertakings if they were well-financed and proceeded with proper caution must have been sustained by the work of two of his assistants, Othmar H. Ammann and David Steinman, two of the greatest engineers of their era. Ammann actually went on to wrest the position of chief engineer on the George Washington Bridge from Lindenthal. Spanning the Hudson was always Lindenthal's fondest dream, but his sensitive political record as bridge commissioner always stood in the way.

The Hell Gate bridge is the essence of Lindenthal's aesthetic conception of bridges. It is not "beautifed" with decoration. Instead, its beauty lies in its very profile—the gentle, refined curve of the arch. In most overhead arch bridges, the upper and lower arcs are parallel in a parabolic curve and both intersect the suspended floor. However, Lindenthal designed the upper arc of the Hell Gate Bridge with a slightly reversed curve, softening the appearance of

COURTESY OF THE NEW YORK PUBLIC LIBRARY

Separated abutments are also featured in drawing by architect Henry Hornbostel, who often teamed with engineer Lindenthal.

the arch and making it more visually appealing than the usual geometric type. The reverse curve and the Roman arched granite-faced towers are what produce such an effective skyline. In fact, Lindenthal's departure from the usual stark steel arch so impressed John Bradfield, who visited America while in the process of designing the Sydney Harbor Bridge in Australia, that Bradfield changed his plan for a cantilever bridge to a steel arch that followed the outline of the Hell Gate.

According to David Steinman, the reverse curve pleases the eye because it helps to emphasize the impression of gigantic horizontal thrust being transmitted from the steelwork of the arch to the massive masonry of the towers. But naturally, as befits the plans of a master engineer, this curve is not a gratuitous architectural effect. Rather it follows naturally from the greater size of the N-shaped trusses necessary at the point where the arch abuts the towers, and it allows the correct overhead clearance for the passage of huge locomotives through the 250 foot high towers. The reverse curve also aids in rigidity by allowing a deeper stiffening truss.

The 1,017-foot long arch, of course, is only the keystone in a system of water and land crossings that in itself was the largest of its type in the world. The whole length of the structure, including the arch and approaches from the abutment on Long Island to the abutment in the Bronx is 17,000 feet, or considerably more than three miles. Within these three miles are two smaller bridges: Across the Little Hell Gate the engineer threw up an inverted bowstring truss bridge in four spans, each of 300 feet, resting on two tower piers at either end and three arched piers in the river. Massive concrete towers 155 feet high stand at each end of this bridge, and again the towers were artistically designed. The Bronx Kill was to be crossed by a trunnion bascule bridge with 175-foot leaves, but a fixed span was built instead. The caissons for the piers of this bridge had to be sunk to a depth of 90 feet to reach solid rock.

But it is in the grandiose steel arch that the ingenuity of Gustav Lindenthal really showed itself. David Steinman, in *Bridges and Their Builders,* says that Lindenthal could have erected a suspension bridge or cantilever bridge over Hell Gate at less expense, and there is even a sketch of a suspension bridge in an early prospectus for the New York Connecting Railroad. However, Steinman says, Lindenthal chose the arch because he felt it would form the most impressive gateway to the Long Island Sound entrance of New York Harbor. Other sources, including *Scientific American and Engineering Magazine,* attribute the choice of the arch bridge to the sweeping curve of the entire viaduct system. In any case, the building of this arch was marked by an inventiveness and originality in meeting problems unrivaled since the Roeblings.

Because of the grace of Lindenthal's arch, few people, even passing as close to it as the Triborough Bridge, realize the colossal size of the various parts. No existing large span bridge had ever been designed with such huge parts. Some of the steel members were more than twice as heavy and bulky as any parts ever hoisted in previous bridge construction. One account

As early as 1903 plans for an 840-foot suspension bridge were published.

COURTESY OF THE NEW YORK MUNICIPAL REFERENCE LIBRARY

COURTESY OF THE BROOKLYN PUBLIC LIBRARY BROOKLYN COLLECTION

Inverted bow string truss bridge crosses the Little Hell Gate. Its four concrete towers are 155 feet high.

COURTESY OF THE NEW YORK PUBLIC LIBRARY

Bundled up against the winter wind, an inspection party halts on the bridge for Dedication Ceremonies. Gustav Lindenthal is the bearded gentleman furthest right on the observation car platform. Date is March 10, 1917.

of the stupendous arch girders maintained that "Were it not for webs and braces running throughout these it would be possible to drive a load of hay through them." The heaviest bottom chord sections weighed 185 tons. They were made of a recently developed material—carbon steel—which gave greater strength for its weight. The bridge, it was said, used more steel than the Manhattan and Queensboro Bridges combined.

Such a heavy structure had to rest on absolutely solid rock, which was found a mere 30 feet beneath the surface on the Queens side. At the Wards Island end, however, there were problems. Hell Gate was always a menace to navigation and is still a difficult channel to negotiate due to its conflicting tides. It was known to be underlaid with rock strata uptilted in a way that might make caisson work difficult. A gas tunnel that had been run across the river nearby had encountered a fissure, which might endanger the bridge's foundations. Extensive borings revealed no breaks in the bedrock. The rock lay at a considerable depth and was very irregular in profile. Lindenthal told his assistants to use a number of small rectangular caissons joined together and surrounded by rows of cylindrical caissons, instead of the usual large boxes. All were to be connected at the top with a huge slab of concrete. The sandhogs worked for several weeks, but one of the cylindrical caissons failed to touch rock, even when lowered down far below the estimated depth. This was proof that the fissure extended into the foundation site of the bridge after all. As the other caissons were carried down it soon became evident that the cleft in the rock extended diagonally across the foundation. This presented quite a problem. Such a weakening of the foundation of an arch bridge would of course be dangerous. However, moving the foundation at this stage was too expensive to be considered.

The solution was simple. The resourceful engineer decided the underground chasm could be bridged just as one on the surface could. He threw a concrete arch across the fissure where it passed through the center of one of his rectangular caissons. At another point, where the fissure lay at a joint between two of the caissons, he bridged the gap by means of a concrete cantilever. The idea of building concrete bridges in a caisson was absolutely original and a contribution to engineering technology. Fortunately, the material covering the rock beneath was clay, so it was not necessary to take great precautions against the entrance of water into the caissons.

Once the foundations were secure, Lindenthal could erect the skeleton of the bridge, but here again there were Promethean difficulties. Because Hell

PHOTO BY VICTOR HAND

Gate is the narrowest reach of a heavily traveled waterway, it had to be kept open to navigation all during the construction of the bridge. Thus the builder was unable to use falsework, the conventional technique in arch building. Instead, the arch had to be built by "overhang" from each end, its arms reaching out simultaneously from the main piers until the crowning connection could be put in place. Meanwhile, the steelwork was guyed back onto massive temporary structures built onto the towers. These structures were heavily counterweighted to balance the weight of the incomplete parts of the arch. It was the first time this method had been used. For the sake of economy, the supporting structures were made of rectangular plate steel and were used in the bridge itself once the arch had been completed.

The most powerful hydraulic jacks in existence were used in closing the arch. When the ends met 280 feet above the waters of Hell Gate, an adjustment of only 5/16 of an inch was needed to bring them into line.

Lindenthal's design also introduced the use of special large rivets, heavy splicing, special pivots or rocker joints at chord splices, and breaking girders in the floor. It is a two-hinged arch, though the hinges are hidden by a steel housing close to the pylons. This gives the impression of a hingeless arch. The hinge-type arch is necessary so that the pylons can take some of the thrust of the chords and it is not all delivered to the ground as is the case with a hingeless arch. Despite the illusion that both the upper and lower chords exert the thrust, in truth only the lower chord does so. As originally constructed, the bridge had three hinges, but the crown hinge was eliminated by bolting up a cover plate at the top of the chord to transform it into a riveted joint. A two-hinged arch is stiffer than its three-hinged counterpart.

Four ballasted railroad tracks are suspended from the arch, while walkways on the outside of the four tracks are supported by cantilever braces. These walkways, used for bridge maintenance, are reached by way of stairways within the granite-faced pylons.

At the time of construction, the mighty towers and the extensive chain of masonry arches of the as yet unconnected viaduct led one journalist to write: "With the addition of a few hieroglyphs, the new concrete piers of the approaches to Hell Gate Bridge would be unmistakably Egyptian. They are 75 feet high, and as you look down through the archways, you think you are standing in the portico of a mammoth unfinished temple."

"There is a spirit of adventure in every big bridge. The careful engineer—and no other kind should be allowed to build a bridge—has every detail calculated, but there is the adventure of seeing the plans work out perfectly and the haunting dread that perhaps some detail will fail," said Lindenthal when he completed the Hell Gate viaduct. In reality, nothing about this bridge has failed except the railroad system it was built to serve. It is still the strongest steel arch bridge in the world, and if it only gets some paint may very well stand as long as the Egyptian temples to which it has been compared.

The Bronx Kill has been reduced to a large ditch. At top is the Bronx Kill section of the Hell Gate Bridge with a fixed truss span instead of the planned trunnion bascule with 175 foot leaves. Also shown is a section of the Triborough Bridge with provision for a lift span if needed.

COURTESY OF THE NEW YORK TRANSPORTATION ADMINISTRATION

HARLEM RIVER BRIDGES

A BAKER'S DOZEN

Manhattan was once severed from the mainland by two streams at its northeast and northern limits. On the northeast side of the island the Harlem River flows parallel to the Hudson for about five miles and empties into the turbulent narrows of the Hell Gate. A small stream, Spuyten Duyvil Creek, originally formed a connecting jagged S between the Harlem and Hudson Rivers. This strait was filled in after a Ship Canal was dredged in the late 19th century to connect the Hudson and Harlem. Spuyten Duyvil Creek is believed to be the place Henry Hudson first dropped the Half Moon's anchor in the New World in 1609.

There are several explanations for how the serpentine strait got its peculiar name. The most famous version is from Washington Irving's *Knickerbocker Tales.* New York Governor Peter Stuyvesant dispatched a trumpeter, Anthony Van Corlaer, to warn the villagers to the north that the British were coming to seize the Dutch villages in and around New York. When the trumpeter arrived at the creek it was a stormy night and Van Corlaer was unable to convince anyone to row him across. After working himself into a rage because he could get no help he took a few swigs of liquor from his trusty stone bottle and declared in Dutch that he would swim "en spijt den Duyvil" (in spite of the devil.) He threw himself into the swirling waters and promptly drowned.

Across the narrow waterway is Bronx County, named after Jonas Bronck, purportedly the area's first European settler. The western half of The Bronx was annexed to New York City in 1876, the rest in 1895. Until then, the heap of villages to the north of the river was part of Westchester County.

Today, much of The Bronx is considered a disaster area. Few travel there for nostalgic sightseeing. But less than a century ago, The Bronx was a sylvan delight, rich with legends and landmarks. It was here that James Fenimore Cooper's Mohicans menaced the settlers during the French and Indian War. Mark Twain and Edgar Allan Poe resided temporarily in the semi-rural hills of lower Westchester County. Poe, it is said, often frequented High Bridge in his afternoon walks.

For colonial New Yorkers, the path up Broadway across Spuyten Duyvil was the lifeline to the mainland. At first the settlers on both sides of the river, like the Indians before them, crossed the waterway by boat. But at low water this was hardly necessary. At the point where the Harlem joined Spuyten Duyvil Creek just west of 230th Street and Broadway, it became so shallow at low tide that it could be forded with ease. A town established on the Bronx side of this crossing was appropriately called Fordham, still the name of a large section of The Bronx.

When Johannes Verveelen, a Dutchman, got a franchise and attempted to run a ferry service from 125th Street, Harlem to The Bronx, he made no profit and changed his location to the wading place. Verveelen then attempted to fence off the crossing so the settlers would be forced to use his craft. The fences were invariably ripped down by the settlers, who claimed the right to walk their livestock across the creek. However, Verveelen built a successful inn at the location.

In 1693 the ferry gave way to a wooden bridge. The bridge was called Kingsbridge and was operated by the aristocratic Philipse family, Bronx land barons who owned the property on the Westchester side of the bridge. The bridge franchise stipulated that everyone who crossed had to pay for the privilege except for soldiers and other representatives of the king, hence the name. It was one of the first toll bridges in America and served to remind the surrounding community of the power of entrenched privilege. The fording place was effectively cut off.

The Philipses took over Verveelen's inn at the ferry crossing and hired as manager a fellow named John Cock, who, it was later discovered, was a British spy. James Fenimore Cooper mentions the inn in his novel *Satanstoe.* Hero Corney Littlepage stops at the inn between adventures to have a recuperative drink with a friend.

A second structure, more ambitious than the first, was constructed 20 years later to replace the original Kingsbridge. This second bridge was 24 feet wide with rough stone abutments laid without mortar. The cost probably was a few hundred dollars, but it stood with only minor changes in its wooden superstructure until 1917. Kingsbridge was shunned for several years when Westchester citizens built a free bridge some time before the Revolution. After the war the new government confiscated the Philipse estate and Kingsbridge was made free. Many years later it became a favorite route for Gay Nineties bicyclers heading north to explore The Bronx's scenic hills. It was also the starting point for the second automobile race ever held in the United States. The race, on Memorial Day, 1896, was sponsored by Cosmopolitan Magazine, which offered a $3,000 purse to the victor. However, the wooden structure, which lent its name to an area in The Bronx, could not withstand the strain of World War I auto traffic and was demolished in 1917. Bronx historians suggested the aged bridge be put on display in Van Cortlandt Park, perhaps over Tibbetts Creek, or over dry land, the way London Bridge later was relocated to Arizona to decorate the desert. Instead, it was buried under Marble Hill.

The alternative to Kingsbridge in pre-Revolution days was the Free Bridge, also known as Farmer's Bridge or Dyckman's Bridge. Built in 1758, it was considered one of the most significant revolutionary acts taken by colonists protesting the Tory establishment. One early New York paper, the *New York Gazette,* went so far as to say it was "the first step toward Freedom in this state."

The Kingsbridge tolls levied by the Philipse family cost the average farmer between six and 15 pounds sterling a year to bring his crops and livestock to market. The shoals of Spuyten Duyvil were fenced off so there was no other route. In addition, the bridge was locked and barred at night and travelers had to wait for a bridgekeeper to arrive to remove the barrier. This became intolerable during the French and Indian War. The farmers' goods had been requisitioned by British troops stationed in New York forts. The farmers were generally apolitical, but resented paying the tolls to supply British troops.

In 1758 John Palmer, the merchant who had founded City Island as a rival port to New York, published a declaration which condemned Kingsbridge and urged Westchester residents to subscribe to a free one. The imperious Frederick Philipse, incensed by this display of independence, finagled the British Army into drafting Palmer for service with the British troops fighting in Canada. Palmer hired a mercenary to fight for him and continued with the bridge project. Philipse had Palmer drafted a second time, but again a hired man was sent and Palmer finished the bridge.

COURTESY OF THE NEW YORK PUBLIC LIBRARY

This famous sketch of old wood and stone Kingsbridge appeared in the 1852 edition of Valentine's Manual.

COURTESY OF THE BRONX COUNTY HISTORICAL SOCIETY

Photo of Kingsbridge taken in early 20th century shows little change. Span was demolished in 1917.

When completed, the Free Bridge, standing at 225th St. and Broadway, was two feet wider than Kingsbridge. Its construction was even cruder than that of its neighbor. The pier abutments and retaining walls were of rough dry rubble and the span was of extremely crude wooden beams. When the British routed Washington's troops from New York in the autumn of 1776 the bridge was destroyed, but it was rebuilt immediately after the Revolution ended, and it stood until 1911.

Both bridges had draws to admit small craft. This important factor determined the development of the Harlem River.

After the removal of Verveelen's ferry service from 125th St. in 1669, nothing was done to provide transportation across the eastern part of the Harlem River for over a century. Even if their destination was Boston, travelers had to detour up to Spuyten Duyvil. Finally, in 1774, a wealthy gentleman farmer, Lewis Morris, after whom Morrisania is named, received permission from the State Assembly to build a bridge across the river where the Third Avenue Bridge now stands. The document named a second man, John Sickles, to construct and care for the Manhattan side of the bridge. The work was never accomplished, partly because of the Revolution, in which Morris played an influential role, and perhaps also because the assembly forbade tolls on that bridge.

In 1790 Morris obtained a second franchise from the State Assembly to build a dam bridge from Harlem to Morrisania. This time he was permitted to collect tolls. A direct connection from New York City to Morrisania was part of Morris' scheme to induce his good friend President Washington to make Morrisania the nation's capital. However, by 1791 L'Enfant had submitted his plans for Washington, and Morris no longer was as interested in immediate construction of the bridge. He assigned the task to John B. Coles in 1795.

Coles, in accordance with the act of 1790, constructed a bridge with a stone dam as a foundation. This dam held back the Harlem's waters and furnished power for the mills which were to be established along the riverbanks. Navigation of the stream was not impeded, however, as Coles provided a passage for vessels which was attended at all times by a lock-keeper. Coles was granted the same toll rate as had previously been approved for Morris. As long as he would keep the 24-foot wide span in good repair, the builder could charge tolls ranging from 37½ cents for a four-wheel pleasure carriage and horse to three cents for a pedestrian and one cent for each ox, cow or steer. He retained this privilege for 60 years.

The bridge diverted much traffic from the Kings and Farmers' bridges for eastern travel from New York, and became a financial success. It was so well patronized that in 1808 the owners incorporated as the Harlem Bridge Company. The well-maintained span was the principal artery of travel to Boston and Connecticut. Because the Harlem and Morrisania areas prospered, the bridge company tried, in 1858, to have its charter extended. It waged a vigorous campaign, but the Legislature, noting that the structure was already becoming inadequate, empowered New York and Westchester counties to maintain it or to build a new bridge. That year it became a free bridge, and soon the two counties prepared to build a new, more capacious bridge across the river. The next Third Avenue bridge built was the first iron bridge built in New York. It was built in 1858 but was torn down in the 1890's.

One more bridge was built across the Harlem River during this early period when carpenters and shipwrights designed bridges on a trial-and-error basis. This one was to become a problem as its builders paid no attention to keeping the Harlem River navigable.

When Philipseburgh was forfeited after the Revolution, much of it, including the land surrounding Kingsbridge, was bought by a wealthy, obstreperous Irish merchant named Alexander Macomb. He had made his fortune in fur trading in Detroit. He also held about 100 acres in The Bronx and additional land elsewhere. In fact, Macomb had managed to purchase from New York State about three and a half million acres of land at eight pence an acre. This tract included the Adirondacks, which for many years were known as Macomb's Mountains.

In 1800 Macomb received a grant from the city of New York to divert the waters in Spuyten Duyvil Creek around Kingsbridge in order to run a gristmill on the shore. In the grant a conditional clause called for the maintenance of a 15-foot passageway for small boats. Macomb, who was well connected, considered this provision a mere formality. He did not bother conforming with the regulation despite the grant's provision that the city could repossess Macomb's property if it were not fulfilled. In any event, the venture was entirely unprofitable and Alexander Macomb's property in The Bronx was sold under foreclosure. The buyer, curiously enough, was Macomb's son, Robert.

The Macombs had achieved nationwide prominence as Macomb's other son, Alexander II, one of the first graduates of West Point, was a hero in the War of 1812. In 1813 Robert Macomb obtained permission from the State to erect a dam across the Harlem River from Bussing's Point in Manhattan to Devoe's Point in Westchester. The dam, which stood where today's Macombs Dam Bridge stands at 155th

COURTESY OF THE BRONX COUNTY HISTORICAL SOCIETY

Farmer's Bridge or Dyckman's Bridge was the free bridge built by farmers who refused to pay the tolls on Kingsbridge.

COURTESY OF THE NEW YORK PUBLIC LIBRARY

On the eastern end of the Harlem River, Coles built a wooden bridge with a 24-foot draw.

COURTESY OF THE MUSEUM OF THE CITY OF NEW YORK

"Bass Fishing at Macombs Dam, Harlem River, N.Y." is a Currier and Ives lithograph from the year 1852. The picture was made after a lock was constructed in the dam. High Bridge Aqueduct, completed in 1842, is in the background.

COURTESY OF THE NEW YORK PUBLIC LIBRARY

Coach and four cross Macombs Dam Bridge, bringing tourists to the clear air and sylvan hills of The Bronx.

Street, turned nearly half the Harlem River, up to the Kingsbridge section of Spuyten Duyvil Creek, into a large mill pond. Robert Macomb's grant contained the same proviso as had his father's. Robert was to have a lock or apron or other opening in his dam to permit passage of small boats and to have a lock-keeper in attendance to open the lock and assist boats through. The annual rent for this privilege was $12.50, the same amount his father had paid for damming Kingsbridge. Robert, like his father, blithely ignored the condition that he must keep the river navigable. Over the dam was a bridge on which Macomb charged tolls, with neither Legislative nor New York City authorization. Young Macomb considered the bridge, which is sometimes referred to in history books as a dam, a great public convenience, and claimed he was donating half of all tolls he collected to charity. These charitable contributions stopped after a couple of years, and young Macomb fared as badly with the Bronx Mill as had his father. His property was sold out by the sheriff in 1817 to the New York Hydraulic and Bridge Company, which put forth an elaborate plan for mill sites and a manufacturing village. This company, which took control of the dam, continued to charge tolls and made no attempt to restore the river's navigability.

Finally, in 1838, a group of exasperated citizens, having failed at getting the company to open the bridge through conventional petitions, decided to take dramatic action. They engaged legal counsel and devised an adventurous scheme to bring the matter to the attention of the U. S. Courts. Their argument was to be that neither the state nor the city had the power to grant the privilege secured by Macomb and his successors in the obstruction of a navigable stream, as this power is vested in the United States alone. Because Kingsbridge and Farmers Bridge had always had draws, it was easy to prove that the Harlem River had been a navigable waterway since its discovery by Henry Hudson.

To get their case to the courts, the group chose the youthful but respected Lewis G. Morris (later to become mayor of New York) as their leader, and embarked upon a course of action that parallels a fraternity prank.

After clandestinely constructing Morris Dock about a mile above the site of the present High Bridge, between Macombs Dam Bridge and Kingsbridge, they chartered a schooner, the *Nonpareil,* to deliver coal to the dock. Morris, commanding the *Nonpareil,* brought the craft and cargo up the Harlem, arriving at Macombs Dam Bridge at high tide. Surrounding him was a party of nearly 100 local residents on flatboats, armed with picks and shovels. Morris demanded passage upstream. As there was no lock, the proprietor of the dam was unable to comply with the request. When the bridgekeeper refused, the locals proceeded to chop up the dam until enough had been destroyed to permit the craft's passage.

The furious owners attempted to have Morris indicted as a disturber of the public peace. However, all legal counsel declared that Morris was entitled to demand passage. When the case finally came to trial, the judge charged the jury that the dam was a public nuisance and that anyone had the right to put an end to it.

As a result, no one could ever again consider obstructing the Harlem River with a low stationary bridge, and John Bloomfield Jervis, who had just received approval for his plans for the Croton Aqueduct System, had to revise the design for the bridge bringing water pipes across the river. The result is **High Bridge**, the oldest extant bridge in New York, and one that won international acclaim for its unwilling designer.

Today High Bridge is something of a hybrid. A steel ribbed arch built in 1927 replaced five of the original granite arches. But for over 60 years the series of granite arches marched across the rural Harlem Valley like a tribute to Caesar's legions. The bridge was the subject of many paintings and a favorite haunt of Edgar Allan Poe as well as other, less famous Bronx strollers. Miller's *Strangers Guide for the City of New York,* printed in 1860, recommends High Bridge as an outstanding tourist attraction, well worth the five dollar carriage fare which allowed passengers to remain at the bridge for two or three hours. One 19th century architectural authority maintained that the multiple arch bridge was the one example of public works in the United States that could compare with Europe's.

John Bloomfield Jervis, who had earned his reputation as chief engineer on the Erie Canal, was reluctant to undertake the $900,000 bridge. A strict Calvinist, whose diary displays his pride in building economical structures and serving the municipal government well, Jervis felt High Bridge was a waste of the taxpayers' money. He claimed that the court order which insisted on a high bridge or pipe line beneath the river was due entirely to the machinations of a few landowners who thought their property would be enhanced by the building of an ornamental bridge nearby. Jervis had first proposed a low syphon bridge that followed the profile of the deep Harlem Valley and crossed by means of a low arched bridge that contained a 50-foot draw. This plan was accepted by the New York Water Commission despite the objections of a number of citizens who wanted to build a structure which would lend to New York some of the grandeur of imperial Rome.

After all, the local citizens argued, the Croton Aqueduct System was the greatest such work in modern times and deserved a monumental bridge. The only other reasonably modern water supply system on the continent was in Philadelphia. Not since Roman times had water been carried such a distance. The longest aqueduct of the day, in London, was only 23 miles long, compared with the 41 miles from Croton Reservoir to the one at Murray Hill (now the site of the New York Public Library at 42nd Street and Fifth Avenue). The other major outlet for Croton Water was the reservoir in Central Park near 90th Street. This has been retained for emergencies, but is fed with water from the New Croton Aqueduct.

New York citizens had been agitating for a better water system for over 50 years. Before the Revolution, a yellow fever epidemic had ravaged the city and the citizens decided the city's few wells were impure and inadequate. It was suggested that water be drawn from either the Bronx River on Morrisania Creek in Westchester to a reservoir at Pearl and White Streets. The Revolution postponed these plans. After the war, the city corporation commissioned engineering studies and petitioned the State Legislature for permission to build a reservoir and raise the needed revenue. However, two extremely influential men—Aaron Burr and Alexander Hamilton, oddly enough—convinced the city that it would be easier for a private corporation directed by Burr's bank to raise the necessary capital. The city would receive one-third of the company's stock, but would not participate in its management. Thus, the city parted with power over its water system, entrusting it to the Manhattan Company for a share of the profits. Although this company was privileged by its charter to bring water from any location in Manhattan or Westchester, all it ever succeeded in doing was sinking a large well at the corner of Duane and Cross Streets, in one of the most thickly settled portions of the city. The water the company pumped up was little better than that which came from the rest of the city's wells.

The Manhattan Company claimed more than $1 million in capitalization, but its primary interest was banking. It was so niggardly in supplying water to New York that when a pestilence in the city in 1801 called for the cleaning of the streets, the city had to compensate the Manhattan Company with $750 for cleaning the gutters with water from their reservoir. Although the city tried to convince the Manhattan Company as early as 1804 (under Mayor De Witt Clinton) to cede their works and privilege of supplying water, nothing came of it until 1831, when on the recommendation of the Fire Department, the city took steps to repeal the powers of the company. Over $600,000 in property had been destroyed the preceding year due to difficulties in procuring sufficient water for fire fighting.

Again a commission was set up to explore the feasibility of obtaining pure water from the Bronx River, Morrisania Creek, Rye Pond, or the Croton River. It was decided that, although expensive, the Croton source was best, as it would supply the citizens of New York with pure water for years to come, furnishing over 40 million gallons a day. The Croton Reservoir also was situated at a high level, so that it could supply the upper floors of New York buildings. (This, in fact, was quite an improvement over Roman aqueducts, which, according to Lewis Mumford in *The City in History,* never supplied water to any floor above the main level.) On June 2, 1835, Major D. B. Douglas, Esq. was appointed chief engineer. Douglas' work, while technically proficient, proceeded very slowly. There were great difficulties in reconciling the inhabitants of Westchester to the fact that they had to part with some of their land, albeit for a generous sum. The farmers, wary of the noise and bother the construction would cause, demanded that in addition to the money the city would pay, they should retain the use of their lands and get free water.

The Water Commission suspected the delays were due to corruption on Douglas' part. They dismissed him quietly, without a reprimand, and hired Jervis. Jervis was a self-taught Horatio Alger type who had come to prominence during the infancy of civil engineering in this country. A neighbor who was an engineer on the Erie Canal had hired Jervis as an axman on the project. Noting the boy's intelligence and talent, he began to train Jervis to do increasingly more difficult work on the canal. Finally, when the canal was improved in 1825 Jervis was named chief engineer. Port Jervis, New York, is named for him. Jervis resided there while designing the D & H Canal. During the building of the Croton Aqueduct System, Jervis took part in forming the American Engineering Society. Jervis is also known to railroad buffs as an innovator in early locomotive design, introducing the first successful four-wheel pilot truck in 1832.

The plans Jervis received from Douglas for the aqueduct were somewhat vague, according to Jervis' diary, despite the fact that the geographical lay of the aqueduct route had been surveyed twice. Jervis considered Douglas irresponsible, especially in his plans to cross the Harlem River Valley with a monumental high bridge. "There is no work under contract precisely similar, or of the same magnitude, or which, from its elevation and inconvenience of access will be so expensive in laying up or require so great a portion of large stones or the same exactness of execution," complained Jervis to the Water Commisson. Jervis defended his plans for a low syphon bridge in terms of cost, experience in sinking the proper kind of piers and the shorter time span needed to bring it to completion. The Water Commission agreed with him and advertised for contractors to do the work on the bridge.

However, the opposition wanted the lofty aqueduct bridge which would run across the valley on a continuous grade, and they were not willing to let the Water Commission have its way. Ads were placed in papers threatening contractors that they would receive no money if they did the work on the low bridge. Finally, after Morris won his case in court, the high bridge faction petitioned the State Legislature to prevent the building of the low bridge. The legislature passed a law requiring the aqueduct to either pass beneath the river by means of pipes, or to be placed on a structure which allowed spans of at least 80 feet across the river at high tide.

COURTESY OF THE NEW YORK PUBLIC LIBRARY

Lithograph shows High Bridge during the construction of a water main. View is from the west gate house.

COURTESY OF THE BRONX COUNTY HISTORICAL SOCIETY

Photo of High Bridge from 1908 shows aqueduct with original stone arches. Washington Bridge is in background.

Jervis examined both methods of conducting the water, and although the tunnel appeared the less expensive venture, decided that not enough was known about tunneling to assure that the estimate was accurate. The Water Commission accepted Jervis' reluctant recommendation for a high bridge, and appropriated the $950,000 necessary for a 1450 foot granite viaduct stretching from one side of the valley to the other. The original design consisted of 15 circular arches of which eight had the requisite span of 80 feet, while each of seven land spans had a 50-foot span. The entire bridge had a clear height of 114 feet above high tide.

While the long viaduct looked almost exactly like the stone ones constructed throughout the European landscape during Roman rule, the 19th century counterpart has hidden innovations. The outstanding one, not readily observed, was even more advanced than those stone bridges being built concurrently in France which used the most modern techniques. The loads from above the arch ring (actually an arch slab) were made hollow, having only the material necessary for strength. Passages were provided from the spandrel walls to the hollow space in the piers to allow water which might fall between the parapets to exit into an opening in the pier near the high water line of the river. This hollow space between the sidewalls of the arch reduced the dead weight. This innovation was copied several decades later in the Memorial Bridge in Washington, D. C. Although High Bridge was not the longest stone arch in the world, even at its time of construction, its tall, slender piers and great length (almost 600 feet longer than the famous Pont du Gard) won it international acclaim.

Two 33-inch pipes were laid within the arch walls to conduct the water. They were covered with five feet of earth to withstand the influence of frost and heat at such an exposed elevation. In 1906 a third, wider pipe was added and the floorway of the bridge was raised to accommodate it. There are also two gate chambers which let the water into and out of the pipes on the bridge.

High Bridge was certainly the most impressive single element in the Old Croton Aqueduct System, but its walkway was not finished in time for either of the two municipal celebrations inaugurating the new water supply. The first took place on July 4, 1842, when, along with the usual Independence Day revelries, the Water Commission gave notice that the first water was to be introduced at sunrise into the Murray Hill distributing reservoir and that all citizens and strangers to the city were invited to examine the phenomenon. A 41-gun salute was fired—one for each mile between the Croton River and the distributing reservoir. At precisely sunrise there was a single gun shot and the gate was raised and the water poured with a bounding rush into the great basin. Temperance was the greatest moral issue of the day, aside from Abolition, and the officers of many of the city's myriad Temperance associations attended the dawn ceremony. Mayor Morris and the City Council did not show up until the late hour of 6:00 A. M., the Tribune reported. All newspapers concurred that the day was marked by a decrease in the consumption of the "distilled damnation" which so often ruins civic celebrations.

A second official celebration took place on October 14, 1842. It also paid homage to the ideals of Temperance and Abolition. The Temperance societies were out in full regalia. The Cold Spring Temperance Benevolent Society marched under the banner: "Turn; drink of the pure fountain of life, come with us and be free." Along with more than 30 societies of teetotalers marched trade unions, civic societies, military societies, ladies auxiliaries and bands of every description. The entire parade was eight miles long and was called the greatest civic display New York City had ever seen. The Erie Canal Celebration in 1825 had been the cause of some municipal festivities, but the Croton Aqueduct was entirely a New York City endeavor, and the 280,000 citizens of New York were delirious with municipal pride. Only the Romans could compete with such a magnificent public work, they were reminded by Mayor Morris, but Rome's aqueducts were "cemented by the blood of slavery, while the Croton Aqueduct was due to free men voting to construct at their own cost a noble aqueduct built by free men." The anti-slavery element cheered.

The day's speeches and entertainment for the masses were followed by a private soiree in the mayor's office, which completely shunned wines and spirits, to the delight of Temperance officials. Beverages consisted of Croton water and lemonade.

Mayor Morris praised the people of New York for having undertaken an enterprise "in a style and on a scale greatly beyond their actual or any near future wants, but which, designated to endure for ages, would bear record to these ages, however distant, of a race of men who were content to incur present burdens for the benefit of a posterity they could never know."

Little did the mayor or any one else there suspect that within his lifetime a third and larger pipe would have to be laid on the bridge to supply even more water for the mushrooming population. By the early 20th century the New Croton Aqueduct System and the Catskill Reservoir System were constructed. At the beginning of U. S. involvement in World War I, the High Bridge Aqueduct was shut down (replaced by a tunnel deep under the Harlem River) and in the 1920s the city suggested demolishing the old bridge. The closing of the High Bridge Aqueduct was due to security fever accompanying the declaration of World War I, rather than any inherent fault in the system. The old aqueduct was shut down on February 3, 1917, the same day the German Ambassador was sent back to his country. At the time, three aqueducts and a pipe line were bringing water into the city, and the government felt it was necessary to patrol the systems to prevent sabotage. It was believed that German infiltrators might attempt to flood the city by bombing the aqueducts. As it was easier to patrol two aqueducts than four, the Old Croton Aqueduct and the Kensico Pipe Line were shut down.

As early as 1911, rumblings had been heard from the U.S. Army Corps of Engineers that the High Bridge arches were too narrow and were obstructing the navigation of large craft. A few years later some Bronx landowners and businessmen complained of the obstruction. A report presented to the city authorities in 1915 balked at the demolition or serious mutilation of High Bridge. The paper defended it on the grounds that it was an aqueduct rather than a bridge and only incidentally carried a footpath, and that it was one of the most notable structures in or around New York City. Even the removal of one or more of the piers was deemed to be a serious public misfortune. Apparently the War Department did not press the matter for some time and it was not until the early 1920's that the issue was brought to a head. The War Department then served notice on New York City authorities, declaring that "High Bridge is an unreasonable obstruction to the free navigation of said river on account of insufficient clearance between piers, and it is proposed to require the following changes to be made in the bridge within one year...: Two alternate piers to be removed and a vertical clearance of at least 100 feet above mean low water to be provided in each of the proposed widened spans."

As there was no longer water flowing through the pipes of the Old Croton Aqueduct, the Commissioner of Plant and Structures and the Board of Estimate concurred that it would be more expedient to demolish High Bridge than to remodel it. They reported that if piers were removed from High Bridge, maximum clearance would remain at 100 feet, still limiting the size of ships using the Harlem River, and that there was no water passing over the bridge anyway.

Ordinary citizens and many professional organizations protested the decision. The American Institute of Consulting Engineers, the New York Chapter of the American Institute of Architects, the American Society of Civil Engineers and the American Institute of Fine Arts all favored preserving the bridge. *Scientific American* announced that the destruction of High Bridge would be "an act of vandalism... without a precedent in the history of our country."

For the next four years debate continued. Various designs were submitted calling for knocking out several piers and substituting piers which would blend with the original design. One faction wanted to go back to a late 19th century plan to widen the pedestrian roadway for automobiles. Both plans were considered too expensive. Finally, in 1927, a compromise was reached when the city accepted a plan to remove four river piers and five arches of the bridge and substitute a steel arch. This change gave the present lateral clearance of 360 feet. The cost for the revision was slightly over $1 million and the city's boosters claimed the reprieve for High Bridge

COURTESY OF THE NEW YORK TRANSPORTATION ADMINISTRATION

With 5 original arches replaced by a steel one, High Bridge resembles her two neighbors—the Alexander Hamilton and Washington.

actually would earn money for the city in the long run. "High Bridge, the Hell Gate Bridge, Brooklyn Bridge and Washington Bridge are magnets which bring sightseers and money spenders to New York City. Whatever can be done to promote this beauty of New York will promote its money-making power," noted one journalist. He also pointed out that a European city would never demolish such a charming and unique ornament.

That journalist, and others who were involved in the revision of High Bridge, could never have predicted the sad condition of the stone bridge today. Although it was designated a landmark by the New York Landmarks Preservation Commission in 1970, the lofty structure is now fenced off with tangles of barbed wire reminiscent of the Berlin Wall. The Parks Department, which has jurisdiction over the bridge, claims that when High Bridge was open to pedestrians malicious youngsters from the surrounding neighborhoods used to break chunks of crumbling stone from the deteriorating structure and drop them on passing excursion boats. The Parks Department does not intend to open the bridge, which connects High Bridge Park with a smaller park on the Bronx side of the Harlem Valley, because it says it has no funds to patrol either the parks or the bridge and it considers the entire area to be entirely too hazardous to be a tourist attraction. The Water Commission notes that the 90-inch pipe inside High Bridge is still capable of functioning in an emergency, but has not been used in years.

Several other bridges were built over the Harlem River before the era of modern bridges. One was a substitute for the original Macombs Dam Bridge, the object of so much complaint. The second Macombs Dam Bridge was undertaken by Jordan L. Mott, the founder of the first successful industrial community in Westchester. A good businessman, Mott managed to get an adequate wooden drawbridge constructed for $90,000. It opened from a high center pier which drew up both ends of the bridge. It served well until 1894, when the modern Macombs Dam Bridge was completed.

In 1858 the cities of New York and Westchester agreed to rebuild the Third Avenue Bridge to provide greater carrying capacity for the two growing districts of Morrisania and Harlem. The railroad craze had brought a number of lines into Westchester, and the population had grown considerably. The city officers, authorized by the state to build a new bridge, concluded that an iron bridge would have the strength to serve the communities far into the future. They hired a French engineer, France at the time being furthest advanced in the field of bridge building and in sinking foundations by the new pneumatic method. He didn't work out well and the city hired one of America's best engineers, John McAlpine, to complete the foundations and design the new iron bridge.

McAlpine was one of the most incredibly hard-working engineers America ever produced. In his 63-year career he was involved in so many difficult and unprecedented jobs that it is hard to imagine any man having enough time for them. His motto, "Integrity, industry, enthusiasm," was taken from that of the newly formed American Society of Civil Engineers, of which he was the third president. It was understood that McAlpine purposely omitted "genius" (found in the society's version) from among these cardinal qualities because that was a gift falling to few and because genius is "proverbially erratic, this quality alone seldom leading to high professional rank."

Genius or not, McAlpine achieved high professional rank. He was the first American admitted into the prestigious French Society of Engineers and to the Institution of Civil Engineers in England. His successes included the laying out of Riverside Drive and Riverside Park, a stint as chief engineer on the Erie Canal after his master John Jervis left, and the practical completion of the U. S. Navy Yard in Brooklyn, then the largest stone dock in the world. The extraordinary difficulties overcome while sinking the Brooklyn dock foundations brought McAlpine to the attention of the engineering world and even the political world. He was elected State Engineer of New York and later became State Railroad Commissioner. McAlpine was also prolific in his contributions to the engineering world. One of his many papers, on the supporting power of piles, won him the Telford gold medal, an English engineering award named after the great English bridge engineer. McAlpine was the first American to receive it.

Although a master technician and a dedicated engineer, McAlpine did not become a rich man. He also lacked the visionary qualities and excitement of a Roebling or a Lindenthal. As the Third Avenue Bridge shows, he was technically competent, but he did not foresee the future requirements of the communities involved. In 1860, McAlpine designed a graceful three-arched iron drawbridge. The piers for the structure were iron cylinders sunk by the plenum pneumatic process, possibly the first example in America of the use of compressed air for sinking foundations. McAlpine was successful in implementing this new technique and he was also able to overcome the difficulties presented by huge boulders found in the path of the cylinders.

However, McAlpine paid absolutely no attention to possible growth of the two neighboring communities. At the very time Roebling was building the Niagara Suspension Bridge to carry America's omnipresent railroad trains, New York City was getting nothing but a narrow iron structure incapable of supporting a trolley. By the 1890's, when elevated lines were being extended out of Manhattan, another bridge had to be constructed at Third Avenue. The $2 million spent on the iron bridge in 1860 proved extravagant.

McAlpine's Third Avenue Bridge marks the first documented instance of corruption in New York bridge building. When costs reached nearly 10 times the $200,000 projected by McAlpine, a committee was formed to report on the circumstances surrounding the New Harlem Bridge, as it was called. The committee completed its book-length report in 1864. It concluded that although New York City (then under the Tweed Ring) was noted for its corruption, the culprits in this case were Westchester politicians who had deceived the taxpayers by insisting upon the iron bridge in the first place in order to place contracts and supervisory jobs in the pockets of their friends. For the investigating commission, Jordan Mott testified, "I have never seen the first man that owned property on the Harlem River or in Westchester that thought we wanted an iron bridge, except those that are concerned in the commission or contracts and I have said I would build them a new bridge and in addition a canal from the North to the East River 10 feet wide for a quarter the amount this has cost." Mott's analysis proved accurate. The second Macombs Dam Bridge, costing less than 1/20 the Third Avenue Bridge, had essentially the same lifespan as its more expensive iron counterpart and served its community equally well.

Also during this period two railroad bridges were built to serve the various lines heading north from New York. One of these was a wooden swing bridge which served the New York and Harlem Railroad, and was located at approximately the same place as today's Penn Central Railroad Bridge at Park Avenue. Another was the New York and Hudson River Railroad's bridge at the western end of Spuyten Duyvil, built on the same site as today's span.

In 1884, a precedent-setting bridge was erected at Madison Avenue. It was the first high-level iron swing bridge on the Harlem and worked on a central pivot pier. The bridge was narrow and delicate. It had to be reconstructed 20 years later to carry increased traffic. But its clearance, when closed, was 24 feet above mean high water, which permitted many boats to pass beneath while the bridge was closed. An efficient steam engine enabled the bridge to open and close for marine traffic in less than three minutes, according to reports. This prompted the U. S. Army Corps of Engineers to set these as minimum requirements for all future drawbridges across the Harlem.

The first really modern bridge erected over the Harlem River is the Washington Bridge, a steel arch bridge located at 181st Street, less than 1,500 feet north of High Bridge. The bridge, which crosses from the heights of Manhattan to high land in The Bronx, consists of two 510-foot steel arches. One arch spans the river; the other accommodates the tracks of the Penn Central Railroad. It is truly one of the most attractive bridges in New York City and was built efficiently, cheaply and well. It has been called the glory of the Harlem River. The idea for the Washington Bridge—initially called the Harlem River Bridge and then the Manhattan Bridge—originated with Andrew Haswell Green, an early Central Park commissioner who helped lay out Washington Heights in the 1870's. Once the streets were laid out and housing was being constructed, Green, an advocate of New York City unity, believed there should be a connecting link which would allow horse drawn traffic to cross from the Heights to The Bronx. The city first considered widening High Bridge for carriages, then decided that a second bridge would be more appropriate. The city immediately surveyed and purchased the land required on either side of the water and had John McAlpine submit several designs for suspension bridges. Other contractors designed plans for cantilever and masonry arches, but no steps were taken to actually build a bridge until the State Legislature, pressured by Green, passed a law in 1885 appointing commissioners for the new bridge.

COURTESY OF THE BRONX COUNTY HISTORICAL SOCIETY

The old Madison Avenue swingbridge ran by steam and set a precedent for the length of time it took a bridge to open and close.

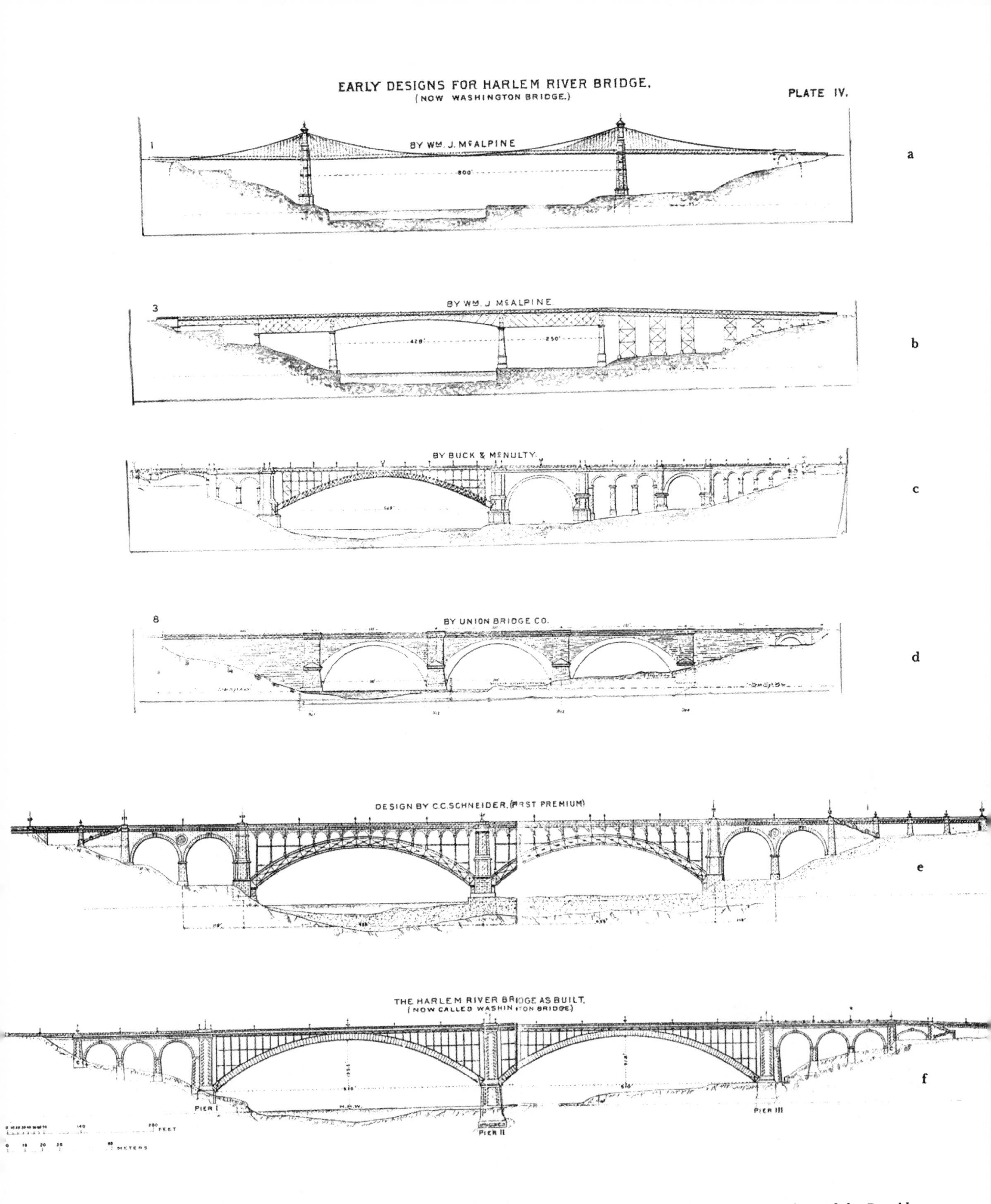

A contest was held for the finest design for the Washington Bridge. Commissioners wanted it to rival the grandeur of the Brooklyn Bridge. Although the commission awarded a prize to C.C. Schneider (for design E), costs for the ornate design were too high, and a compromise was reached with the similar but simpler arch, seen in figure F.

The three new commissioners decided to construct a monument to vie with the new Brooklyn Bridge in dignity and rank in the world of bridges. Its design was obtained through a formal contest, and although the bridge committee awarded the first prize to Mr. C. C. Schneider, they nevertheless were forced to build the bridge according to a less elaborate design in order to reduce costs. Schneider's design had produced a contractor's estimate of $3 million, which brought the wrath of the *Herald Tribune* down on the commission. Thus, they modified the prize winning selection by substituting a steel webwork for the more ornamental iron work the designer had suggested.

McAlpine was chief engineer on the project, but only served in this capacity for about a year, as his health was giving way. He died before the completion of the bridge. His successor, William Hutton, not only did an admirably efficient job, but left a record of the construction of the bridge in a handsome volume published in 1889. *The New York Times* regarded the Washington Bridge as a "marvel of rapidity of construction," for it was completed within two years of its starting date in 1886, though at first it was only opened to privileged pedestrians with special passes.

The approaches to the Washington Bridge are great granite viaducts with masonry arches. Each of its two-hinged arches exceed in length the arches of the famous Eads Bridge across the Mississippi, often considered the most impressive steel arch bridge of the period. Alhough ornamental iron work does not embellish the arches, the lyre-like design along the balustrades and the artistic flourishes on the railings are very European in flavor and call attention to the aesthetic sensitivity of the designer, who wished to please the eye as well as construct a convenience. Plate girders used for the arch ribs give the illusion of stone voussoirs on a masonry arch. It was the first instance in America where plate girders were used for arch ribs.

The *Times* called the bridge "one of the most imposing, beautiful and substantial to be found anywhere about the metropolis and is especially interesting as a perfect and consistent edifice in the arched style of bridge architecture." The original layout for the roadway also reflected a consideration for the human freight to be carried. Two 15-foot walkways were constructed on either side of the bridge. At the center of the roadway, between the traffic lanes, was a grassy mall. Later the roadway was widened, the walkways narrowed and the mall removed.

The bridge, situated at very high bluffs in both The Bronx and Manhattan, offers a panoramic view on clear days. Because of the depth of the Harlem river gorge that surrounds it, however, the bridge gives the illusion of being lower than it is. The Washington Bridge is actually 151.6 feet above mean high water, 16 feet higher than the Brooklyn Bridge, and six inches higher than the Statue of Liberty.

The bridge was adequately completed for pedestrian use by the end of 1888. At that time special

COURTESY OF THE COLLECTION OF PHILIP EAGAN

The Washington Bridge has 2 arches — one over the Harlem River and one over the N.Y. Central R.R.

Scientific American featured this sketch of the Washington Bridge in 1888. COURTESY OF THE NEW YORK PUBLIC LIBRARY

passes were issued to people who wished to walk across. In February 1889 the contractors officially turned the structure over to the commissioners. It was Washington's Birthday, and the bridge was rechristened the Washington Bridge as one of the first moves in the Washington Centennial Celebration. The formal opening ceremonies were intended to add eclat to the national festivities. However, the city fathers decided the winds gusting across the valley in addition to the normally cold weather would discourage crowds, so the official opening was put off until spring. These plans never got off the ground since arguments ensued between the bridge commissioners and the city, and finally public sentiment for such a ceremony dwindled.

Battles over the bridge commissioners' salaries kept the crossing officially closed to the public all year. Finally, in December 1889, the public took matters into its own hands, tore down the barricades and began using the bridge. Consequently, there was no formal opening of the structure; in fact, for a long time after the bridge was in use, signs were displayed notifying all persons who crossed it that they did so at their own risk and that the city disclaimed all liability for accidents.

While the Washington Bridge was being completed, the U.S. Army Corps of Engineers began to widen the waterway from the Harlem River to the Hudson. Spuyten Duyvil had never really been navigable at low tide, and the idea of opening a navigable channel between the Harlem and Hudson dated back to 1827, when a company was formed for that purpose. Work started in 1888 on what was to be called the Harlem River Ship Canal. The new channel measured 350 feet across and is 18 feet deep at mean low water. In 1892 new bridge regulations were formulated. These rules, the spiraling population growth in Bronx County, and the construction of the elevated lines to serve that population caused a spate of bridge building across the Harlem during the next 20 years. Many of these steel swing bridges, which casual views may find indistinguishable, are still standing. Three no longer exist.

The new rulings, as well as the ever-expanding traffic on the New York Central Bridge, forced the railroad to replace its iron bridge with a more capacious steel one. Over 200 trains a day were crossing the old iron bridge by 1889. The original structure, dating back to the early days of the New York and Harlem Railroad, was a two-track swing span that stood only seven feet above mean high water. By the 1860's the pivot pier had been grazed by so many passing ships that engineers doubted its reliability, and the wood swing span of the bridge was replaced by a 90-foot iron draw bridge. But even this was inadequate for the type of marine traffic expected as a result of the newly dredged canal. So the Central constructed a temporary wooden bridge which bypassed the Park Avenue crossing and then built a four track steel swing bridge, completed by 1891, which was one of the first four-track drawbridges ever constructed. The 24-foot height of the new Harlem crossing led to the building of the Park Avenue Viaduct all the way south to 96th Street. Previously the tracks were in tunnels and at grade.

Alfred Pancoast Boller designed the new railroad bridge while working for the Vanderbilt interests. He then established a consulting firm which created four of the city's drawbridges over the Harlem. Boller was considered "intuitively artistic," and in his spare time painted watercolor landscapes. His fondness for the artistic led him to strive for architectural harmony in bridge designs, as exemplified by his Macombs Dam and University Heights bridges.

The city paid half the expenses for the construction of the Park Avenue Railroad Bridge, and the Central the rest. It served both the New York Central and the New Haven Railroads until 1954, when the present lift bridge was completed. A ride on the Hudson Division of the Penn Central, which crosses this bridge and winds its way northwest on the Bronx shore of the Harlem River Valley, affords one of the best views of the Harlem River bridges.

COURTESY OF THE MUSEUM OF THE CITY OF NEW YORK

COURTESY OF THE BRONX COUNTY HISTORICAL SOCIETY

Above: The 1860's saw this iron span built at Park Ave.

Right: It was soon replaced by this 4-tracked swing span.

In 1954 the 4-track swing span was replaced by a lift bridge. It was constructed right alongside the earlier bridge.

Although the Third Avenue and Willis Avenue bridges are not passed, they can be easily seen from the lift bridge, and are architecturally the two least interesting spans over the river. Architectural critics even regard them as eyesores.

The Ship Canal also made necessary the immediate construction of a bridge crossing the new waterway at Broadway. A wooden structure was built for temporary use, and by 1895 Boller finished a single deck steel swing bridge. In the tradition of the Madison Avenue Bridge of 1884, the Broadway Bridge was 26 feet above mean high water and ran on steam, opening and closing within three minutes. Although all the regulations set by the Army Corps of Engineers were satisfied by this bridge, the span proved inadequate within ten years. The elevated portion of the IRT subway had almost reached the canal, and the Broadway Bridge was not strong enough to take the weight of an additional deck for trains. The original span, therefore, was replaced in 1905.

The new double deck swing span was constructed on piles on the shores of the river. When it was finished, four pontoon scows floated its predecessor almost a mile down river to 207th Street. There workmen attached it to a central pivot pier which had been prepared, along with new abutments, to receive the old Broadway span. Dubbed the University Heights Bridge, it still stands with the original swing span from the old Broadway Bridge. Gustav Lindenthal had suggested the University Heights Bridge while he was bridge commissioner in 1903, because there was no direct traffic route from Manhattan to The Bronx in that developing area. Lindenthal, always the innovator, had favored a lift bridge, then a new development in bridge engineering. The city was not overly eager to spend the money on the new, more expensive type of bridge, and when the opportunity was seen to make use of the discarded Broadway span, the city seized it. Since the late 1920's lift bridges have been regarded as more appropriate for the Harlem River, as they give a wider clearance than center pier swing spans, but by saving money here the city did preserve a lovely piece of ornamental steel work. A walk on the University Heights span reveals the aesthetics of the 1890's, when ornament was considered beautiful. Amidst wrecked cars, rotted piers, and oily water, the gay steel cut-outs of cartwheels and daisies lining the walkway give evidence of the playful mind that designed it, as do the green shingled gazebos at either end of the bridge and the peaked silhouette of the span itself.

Two days after the University Heights Bridge received this center span, a 1,600-ton double deck unit was floated into place at Broadway. This operation took 35 minutes. The first elevated train crossed on January 14, 1907. The last train passed over that structure on December 23, 1960, the day before the

COURTESY OF THE BRONX COUNTY HISTORICAL SOCIETY

This early swing bridge at Broadway was soon replaced by a heavier two-level span. This span was then floated to University Heights, where it still serves as an automobile crossing. It was designed by Boller and is studded with ornamental ironwork.

Two-level swingbridge on left began service in 1907. It was discarded when the lift span, below, was put in its place on Broadway on Christmas Day, 1960.

COURTESY OF THE NEW YORK TRANSPORTATION ADMINISTRATION

time-worn span was floated down to the Bronx side to be cut up into scrap. The present 2,500-ton double deck lift bridge had been previously fabricated on barges along the east bank of the Harlem, just north of University Heights. This new 304-foot span was floated into place at high tide about 3:00 p.m. on Christmas Sunday, 1960.

The third bridge to go up after the Ship Canal was completed is the oldest swing bridge still standing in its original form over the Harlem River. It is the Macombs Dam Bridge, completed in 1895. It, too, was designed by Boller with his usual concern for aesthetic effect.

The bridge has been humorously likened to a raffish tiara. Following the precedent of the currently novel Tower Bridge in London, which had used Tudor style architecture in its abutments, Boller had designed Gothic-revival strucures to abut the Macombs swing span. This choice of architectural style makes some sense, as drawbridges did originate with the moats surrounding medieval castles. The Municipal Engineer's Society of the City of New York applauded Boller's bridge for its fine architectural effects: "the fine lines of the masonry and the graceful sweep of the upper chord of the draw span and the red tile roofs of the four little houses, which we consider spots of warm color which relieve the severity." Although today this bridge seems like a flimsy structure, at the time it was built, it was one of the heaviest drawbridges ever constructed. Its location between Yankee Stadium and the old Polo Grounds in Manhattan helped make it the most walked-over bridge crossing the Harlem. It is the only movable bridge over the Harlem aside from the University Heights Bridge which warrants a walking tour. The others, also completed around the turn of the century, are less interesting, or are downright blights on the river.

The earliest of these uninspired works was the Third Avenue Bridge, built in 1898 to replace McAlpine's Third Avenue iron bridge. The River and Harbor Acts of 1892 had served to make McAlpine's bridge obsolete, as its draw was only 80 feet long and it stood only 13 feet above the river. Its successor, designed by T. C. Clark, uses masses of masonry ramps to approach the 24-ft. high bridge. The superstructure is a riveted lattice shaped like a bow-string or bracket turned on its side. Although some ornament existed on the superstructure at the turn of the century, this was stripped when the bridge was modernized in 1953. The original bridge had two parallel trusses running along the center of the bridge, which marked off a separate lane for trolley cars. After trolleys stopped running up to The Bronx, this lane was converted for use by motor vehicles and eventually the inside trusses, which in themselves constituted a complete bridge, were sold by the Terry Contracting Co. In February 1955 the following advertisement appeared in the Public Notices column of the *New York Herald Tribune*: "Wanna buy a bridge? Steel swing bridge, 302 feet long, sixty three feet wide, can carry 100,000 vehicles, 500,000 pedestrians or 10 million chickens daily. Terrific bargain! Cash and carry, or will install anywhere in the world. T 87 Herald Tribune."

This bridge cut the time necessary to open its draw for river traffic to one and a half minutes. Instead of the steam engines used to open the previously built draws, electricity was used. In time, all the bridges were converted to electric power.

According to Livingston Schuyler, and anyone who cares to take a look, the absolute ugliness of the Third Avenue Bridge was exceeded by its nearest neighbor to the east. The Willis Avenue Bridge, completed in 1901, was considered a disgrace to New York. It combined the unattractive "bow string" superstructure of the Third Avenue Bridge with a second span curved with a truss. Schuyler's comment was that the Willis Avenue Bridge "derives an adventitious ugliness from the fact that the curved draw span is flanked, on one side, by a bowstring girder, which would be sufficiently awkward with the central feature if it were symmetrized by being repeated on the other side, whereas it is in fact counterparted by the plate girder of the roadway. One says, with confidence, that the arrangement would be intolerable to a designer of any aesthetic sensibility and that such a designer would find some way of circumventing its awkwardness."

The two remaining swing spans across the Harlem are the 145th Street and the Madison Avenue bridges, completed in 1905 and 1910, respectively. They were also designed by A. P. Boller, and in some respects display his knack for building appealing structures. Both geographically and aesthetically, these two bridges form a middle ground between Boller's earlier structures and the Clark bridges to the east. The 145th Street Bridge attempts to recapture the grace and originality of the Macombs Dam Bridge, but it is an uninspired copy, even if the designer was the same. When a design achieves a certain quality, the lack of challenge in repeating it can give its double a humdrum appearance, which is probably what happened with the 145th Street Bridge. It is, however, a pleasant enough bridge with its green tiled gazebos. For some reason, the Department of Highways has seen fit to paint the bridge bright blue, which also detracts from its appearance. The second Madison Avenue Bridge again reveals Boller's penchant for ornamental iron work. It was originally built with a spire atop each girder.

Several railroad bridges have been built over the Harlem, but the number has diminished. One at Second Avenue was built by the Suburban Rapid Transit Co., which owned the Third Avenue Elevated. This swing bridge was constructed in 1878. It

COURTESY OF THE BROOKLYN MUSEUM

Edward Hopper's painting of Macombs Dam Bridge is a tribute to engineer Alfred Pancoast Boller's fine structure on the Harlem.

COURTESY OF THE BRONX COUNTY HISTORICAL SOCIETY

Photograph shows overwater swing section of the Macombs Dam Bridge, completed in 1895.

COURTESY OF THE BRONX COUNTY HISTORICAL SOCIETY

The Willis Ave. Bridge was called the least attractive on the river. It combines a curved span with a bowstring girder.

was replaced in 1915 with a rare double decked swing bridge carrying two tracks on each deck. Another, the Putnam Bridge at 8th Avenue, was built in 1881 for the New York City and Northern and was later converted for use as a spur for the 9th Avenue Elevated to the Jerome Avenue Subway. It was also a Boller creation. The two bridges had provisions for pedestrians as well as trains. They were torn down in the 1950's with the demise of the El Systems. A third bridge spans the mouth of the Harlem at its junction with the Hudson River. As only a few freight trains a day cross this structure it is left open at all other times. The first bridge here was built in 1840, and its basic design has changed very little since then.

One last monumental bridge was planned during these years. Boller, having succeeded in pleasing most critics with his Harlem River spans, was asked to submit a design for an arch bridge to link the Northwest tip of Manhattan with Riverdale in The Bronx. The city government intended that the bridge be completed in time for the Henry Hudson tercentennial celebration in 1909. Boller's design called for an arcade of Roman triumphal arches on either side of a steel arch. According to Livingston Schuyler, Boller's arcade would rival those at Segovia or Nimes, or any other great Roman work. The arches, Schuyler contended, "do admirably and completely fulfill their purpose of giving 'load' and emphasis to his abutments and thus a visible assurance of the sufficiency to his construction. There will at least be a general agreement that the conception of the Hendrik Hudson Memorial Bridge is noble and adequate and well deserving of execution."

Boller's design pleased the city fathers, who, anticipating the additional excitement the opening of such a bridge would bring to the tercentennial celebration, immediately appropriated $1 million for the new public work. Naturally, the people in the surrounding areas of Inwood and Riverdale assumed that the proposed bridge would be built. Picture postcards were made of the impressive structure from the designer's sketches. Paintings were done. Boller submitted several alternative sketches, all including the triumphal arches which had found favor in designs for the Washington Memorial Bridge in Washington, D. C. One design provided for a central concrete arch, which would have given New York the longest concrete arch in the world.

Despite public confidence that the bridge would be built, the Art Commission vetoed the Boller design. It issued no official statement, but sympathizers conclusions excused the decision on the grounds that Roman arches were hackneyed and not appropriate for the wild and romantic geographic location. Most participants in the skirmish were thoroughly annoyed, for the veto prevented any bridge from being constructed at that location in time for the Hudson celebration. Not until the 1930's did construction begin on the Henry Hudson Bridge. It was built under the jurisdiction of the Triborough Bridge Authority by David Steinman, and will be treated in the chapter on the Triborough Bridge Authority bridges.

COURTESY OF THE AUGUST ROCHLITZ COLLECTION

Swingbridge links 145th St. in Manhattan with 149th St. in The Bronx. It is similar to Macombs Dam Bridge, also by Boller. The schooner from Dover, N.H., at right, is a reminder that the Harlem was once a busy waterway. Photo was taken in 1908.

COURTESY OF THE BRONX COUNTY HISTORICAL SOCIETY

Boller's last swingbridge over the Harlem was built at Madison Ave. in 1910.

Commuter train comes through the four-track Park Avenue lift bridge.

PHOTO BY VICTOR HAND

All movable bridges built over the Harlem River since 1910 have been vertical lift bridges because these afford a wider clearance for passing craft. The easternmost crossing was the largest lift bridge in the world when constructed, and is part of the Triborough Bridge complex. The other two lift bridges, at Park Avenue and Broadway, have been discussed above.

The last structure built over the Harlem was the Alexander Hamilton Bridge, completed in 1963. It is a high-level fixed concrete arch, and is part of the Cross Bronx Expressway route to the George Washington Bridge. The Alexander Hamilton is indicative of today's indifference to human scale: since the new bridge is considered part of a highway, there are naturally no footwalks.

The modern Harlem River Bridges have inspired little literature. During the depression, however, Harry Roskolenko, a poet with left-wing sympathies and later a *New York Times* Far East correspondent, described his job as substitute drawbridge operator on the seven city drawbridges over the river (from the east they are the Willis Avenue, Third Avenue, Madison Avenue, 145th Street, Macombs Dam, University Heights and Broadway bridges.)

"It was like a stationary ship, heavy with switches, with huge circling gears and great turntables thick with grease and oil," he wrote in his autobiographical *When I was Last on Cherry Street.* He described himself among the ordinary Tammany crewmen writing poems to scows and tugs and cantilever arches. Although descriptions of most of the bridges at the time of construction boasted they could open within three minutes, Roskolenko estimated it took four full minutes for the bridges to open and five minutes for them to close. Motorists who have been delayed by the movement of these bridges would probably side with Roskolenko. Records show that the average time to open and close one of the swingbridges is about 9 minutes.

Although the Harlem is spanned by 14 bridges within its five miles, and a walking tour is physically possible, it is not a recommended journey. The bridges, especially at the eastern side of the river, are surrounded by impersonal housing projects, industrial sections, and expressways which are at best unpleasant and at worst quite dangerous. The same is sometimes true for the parks which adjoin High Bridge and the Washington Bridge. Two safer ways of examining these structures are from a train on the Hudson division of the Penn Central, as previously mentioned, or from a boat below. The Circle Line excursion boats are more reliable for bridge watching than most private craft, as the currents of the Harlem are notoriously tricky and may require full attention. Another alternative for courageous bridge explorers is the bicycle.

COURTESY OF THE BROOKLYN PUBLIC LIBRARY BROOKLYN COLLECTION

Completed cables swoop down into unfinished Manhattan anchorage. Tower photo shot Dec. 1930.

GEORGE WASHINGTON BRIDGE

CONCEIVED IN DETROIT

The George Washington Bridge bestrides the Hudson River. Its great naked geometry is set off like a fine jewel by the stately Palisades. Its short side spans place emphasis on the enormous steel towers. Like the Brooklyn Bridge, the George Washington Bridge represents a leap in engineering technology. The Hudson span did not merely set a record for length; its 3,500 feet doubled the existing suspension bridge record when completed in 1931.

The grandeur of this bridge evoked this unabashed tribute from French architect Charles Edouard Jeanneret (Le Corbusier) in *When the Cathedrals were White*:

"The George Washington Bridge over the Hudson is the most beautiful bridge in the world. Made of cables and steel beams, it gleams in the sky like a reversed arch. It is blessed. It is the only seat of grace in the disordered city. It is painted an aluminum color, and between water and sky, you see nothing but the bent chord supported by two steel towers. When your car moves up the ramp the two towers rise so high that it brings you happiness; their structure is so pure, so resolute, so regular, that here, finally, steel architecture seems to laugh. The car reaches an unexpectedly wide apron; the second tower is very far away; innumerable vertical cables, gleaming against the sky, are suspended from the magisterial curve which swings down and then up. The rose-colored towers of New York appear, a vision whose harshness is mitigated by distance."

A bridge derives some of its attraction from its setting. The Palisades of New Jersey, the promontory of historic Washington Heights and the mighty Hudson glistening below frame the beauty of this great bridge.

The Hudson River, also referred to in the early days of this country as the North River (to distinguish it from the Delaware or South River), probably was the most heavily-traveled waterway in 19th century America. After the building of the Erie Canal, raw materials and manufactured items could be brought along the canal from the Great Lakes region to Albany. Albany, in fact, was almost as busy a port as New York City, and in the 1830's more than 500 vessels a month left the state capital bound for ports from Valparaiso to Rangoon. If the Hudson was not the most commercially important U.S. river, it rivaled all others. Its beauty and the rocky precipices along its banks inspired artists to compare it to the Rhine.

It is no wonder that this deep watery cleft stirred the imagination of the best bridge builders, and the history behind the Hudson River span at Washington Heights is an epic in itself. More than 60 years elapsed between the authorization of the first bridge company by the State of New Jersey and the beginning of construction of the George Washington Bridge by the Port Authority in 1928. At least four locations were proposed. Six bridge companies and the state and Federal governments were involved at various times. Eighteen designs were submitted. Some passed the strictures of the Army Corps of Engineers. Some did not. Most were expensive. Most were railroad schemes. It was said that every engineer on the continent had a pet design for the Hudson River Bridge in his vest pocket. To some, like Lindenthal, building a bridge for the North River became a driving ambition. Others fooled with it in their spare time.

In 1868, while Roebling was at work on the Brooklyn Bridge, New Jersey decided that if it were to compete with Brooklyn it would also need a bridge to New York City. That year the New Jersey State Legislature authorized the New York and New Jersey Bridge Company to construct a span over the Hudson at a point near the southern border of Union Township. The proposed bridge was permitted to have one or two piers in the river, but had to maintain an unobstructed span of 1,000 feet or more for the passage of river craft. Overhead clearance was to be 135 feet, the same as for the Brooklyn Bridge. However, the bridge company could not become a reality until New York passed similar legislation, as the Hudson is an interstate waterway. New York waited until 1890 to do so, and then it stipulated that the bridge could not have a pier in the river. Two years had passed since the first bridge over the Hudson south of Albany had been built by railroad interests at Poughkeepsie.

The joint New York and New Jersey Bridge Company applied to the Federal Government for permission to construct a cantilever railroad bridge with a center span of 2,300 feet between the towers. The New Jersey tower would reach 900 feet beyond the pier-head line. The New York tower would be within the pier-head line. The proposed site of the six track cantilever bridge was 70th Street in Manhattan. The bridge companies argued that a "suspension bridge spanning the North River without a pier would involve such elements of uncertainty as regards first cost, novelty in its magnitude as a hitherto untried engineering feat, and time of construction, to say nothing of the well-founded prejudice against the suspension principle for railroad purposes, as would render the enterprise impracticable from a financial standpoint." The cantilever, they said, would cost $22 million.

At the same time, Lindenthal was devising a suspension span for the North River Bridge Company, an alliance of railroad interests eager to bring their terminals from the western shores of the Hudson into the hub of New York's business world.

In 1888, Lindenthal presented his plan to the American Society of Civil Engineers. He suggested it be built in the vicinity of 10th Street. It called for two 525-foot high octagonal steel towers and a 2,850-foot span to be carried by four braced eye bar cables 48 inches in diameter. The design aroused a certain degree of disbelief. Engineers felt that even if the bridge were possible it would never stay within the limits of the $16 million Lindenthal estimated.

The U. S. Legislature and President Benjamin Harrison—whose administration was in abject debt to railroad interests—approved the general project. The Secretary of War was to fix the proper dimensions and stipulations. The President appointed a board consisting of five disinterested bridge engineers to consider "what length of span not less than 2,000 feet would be safe and practicable for a railroad bridge". The Secretary of War appointed a separate board of officers of the U. S. Army Corps of Engineers to determine "the minimum length of span practicable for suspension bridges and consistent with the amount of traffic probably sufficient to warrant the expense of construction".

In an 1890 document that became a classic of engineering research, the boards agreed that a cantilever span of 2,000 feet would be possible, but a single span cantilever of 3,100 feet would double the costs and was not economical. A six track railroad suspension bridge, on the other hand, could easily be expanded to 3,000 feet without prohibitively raising the cost. They also said that the expense of constructing a deep river foundation for a cantilever of even 2,000 feet would balance the extra cost of a 3,100-foot suspension bridge. The extra cost of the deep foundation was due to a deep fissure under the Hudson. They unreservedly recommended the suspension design.

The cantilever plan was scrapped. The bi-state bridge company changed its proposal to a six track suspension bridge with an estimated cost of $25 million. Although a design was prepared and ground breaking ceremonies were held in Riverside Park and borings were made at both 59th Street and 71st Street, no construction work was ever done.

Meanwhile, the North River Bridge Company enthusiastically adopted Lindenthal's design. Although bridge costs were estimated at around $16 million, a New York terminal and the acquisition of land brought the final total to $40 million. As the city's focus of activity was changing, the bridge company moved the site from 10th Street to 23rd Street, Manhattan, to connect with New Jersey terminals in Hoboken. The bridge was not merely to have six tracks recommended by the Army Corps of Engineers, but was to be built with enough strength to hold ten tracks in the future. By 1892 seven million persons and thousands of tons of goods each year were crossing the Hudson by ferry from the various railroad terminals in New Jersey. But despite the urgent demand, the attempt to build the bridge failed, mainly due to a general financial crisis in 1893.

Lindenthal's design was called "bold and well conceived," by Othmar Ammann, the eminent Swiss engineer who eventually executed the design for the George Washington Bridge. In an introduction to *The George Washingon Bridge,* a book-length document put out by the Port of New York Authority in 1932, recounting the construction of the bridge, Ammann wrote that it was completely possible that Lindenthal's bridge (which had undergone several revisions and had been submitted to various interests) might still be constructed. While the George Washington was under construction, Lindenthal had again obtained the sponsorship of the Baltimore & Ohio Railroad and was optimistically trying to renew the North River Bridge Company franchise to build a bridge at 59th Street. But the railroad empires in the 1930's had lost the political and financial power they had had in the 1890's and could not embark on such schemes themselves, and the Hoover Administration was not about to help the B & O construct a $200 million bridge. New York City contributed to defeating Lindenthal's last scheme. It did not relish having another railroad station to further tie up congested city traffic.

This was the last in a series of disappointments for Lindenthal. When the Pennsylvania Railroad, at the height of its power, in 1900, decided to extend its trains into Manhattan by means of a bridge, it had called in Lindenthal to resubmit his Hudson River Bridge design. This time Lindenthal talked in terms of a suspension bridge with 16 tracks on two levels. Not only would it set a record as the longest suspension bridge in the world, but it would also have had the greatest carrying capacity. However, the Pennsylvania wanted other railroad companies to share the financial burden. The other companies agreed to this proposal, but failed to assure their financial support. Alexander J. Cassatt, president of the Pennsylvania Railroad at the time, realized he would get no help from them and decided to build two single track railroad tunnels for exclusive use by the Pennsy. These were completed in 1910. The Hudson and Manhattan Railroad Company in 1908 had completed the first of its two pairs of tubes connecting several of the railroad stations on the New Jersey side of the Hudson with the downtown and midtown areas of New York City. These tunnels obviated the need for a passenger carrying railroad bridge for quite some time. However, many private citizens in northern New Jersey, as well as railroads which didn't have access to the tunnels, loudly advocated the need for an additional water crossing, since an ever increasing amount of freight and automobile traffic was still dependent on the 17 ferry routes which plied their way across the Hudson from the Battery to Tarrytown.

To mollify these groups, New York and New

Gustav Lindenthal first proposed this Hudson River Bridge in 1890. Note his characteristic eyebar bracing between cables.

Lindenthal altered his design and it appeared in Scientific American, *May 1891. It carried 10 railroad tracks.*

Jersey in 1906 appointed public commissions to investigate the best way to cross the Hudson. The New York Interstate Bridge and Tunnel Commission and the New Jersey Bridge and Tunnel Commission were authorized to study a Hudson River vehicular crossing and recommend either a bridge or a tunnel. This was the year the United States, under the brash leadership of President Teddy Roosevelt, was undertaking the Panama Canal, the most spectacular engineering feat of the time. In New York, two huge bridges—the Queensboro and Manhattan—were materializing over the East River, and it looked like advancing technology could not be halted.

The two commissions studied all aspects of such a crossing, giving prime consideration to the economic factors. The two states seemed willing to finance the venture. After four years the commissions recommended the bridge be put at 179th Street since it is there that the river is narrowest, and a deep foundation placed at this point would avoid the deep crevasse which underlays the southern end of the river. The commissions believed a suspension bridge at this location could be built for a mere $10 million, for the height of the cliffs here made graded viaduct approaches unnecessary to meet the 135-foot clearance height set by the Army Corps of Engineers. The investigation had apparently been limited to paperwork, for when borings were made in 1910 at 178th Street to test the bridge commis-

Detail of 525-foot towers displays ornamental steelwork.

sions' hypothesis, it was found that a river pier at this point would meet the same difficulties as any other site on the Hudson.

Lindenthal saw his opportunity and submitted a revised plan for a Hudson River bridge which could carry 16 railroad tracks and would also have several lanes for motor vehicles, for he realized by then that the internal combustion engine was more than a fad. But he still maintained that no bridge could be profitable if it did not carry railroads. His latest bridge would be located at 59th Street. However, the tottering economic situation after the Taft Administration, and a series of anti-trust cases against the railroads during the Roosevelt and Taft Administrations, prevented either the government or the railroad interests from financing the bridge. At this point, Lindenthal, who had found the efficiency and cooperation inherent in working for private railroad interests far superior to his experience with government, wrote: "To insure (the bridge's) success, any plan for connecting Manhattan with New Jersey should include the cooperation of the railroads, at least so far as the handling of freight is concerned. Every one of the four existing East River bridges cost from 80% to 120% more than the original estimate at the time the bridges were authorized, and this is only in keeping with the sad experience which New York City and State have had with regard to other important engineering works that have been built with public funds. If the present grossly unfair treatment of the railroads shall come to an end, that is to say if they are once more permitted to operate according to true economic laws, and therefore should feel justified in facing larger expenditures to improve their systems, they will find the proposed railway connection with Manhattan an attractive proposal." At that time Lindenthal's bridge would have cost $76 million, exclusive of the terminals.

New York and New Jersey were not keen on supplying the money for the bridge. It is possible that if only one state had been involved in the venture, the opportunities for graft and corruption might have tempted some politicians to get a bridge act financed. But the two states intended to watch one another on this project, so the opportunity for personal gain seemed very limited. No legislation was passed. The public was disgusted, as were many politicians who saw the Hudson River Bridge as a necessity in developing the economy of the region. To pacify critics, and perhaps come up with a solution, the states announced a second study. This one was to gather information on the feasibility of a vehicular tunnel as opposed to a bridge. The states reasoned that if this was the solution to which the Pennsylvania and Hudson and Manhattan Railroads had come, it was probably a more economical proposal. The study was made, but no action was authorized. In fact, nothing was done until 1919 when the states instructed the commission to proceed with the construction of a vehicular tunnel between Jersey City and a point in the vicinity of Canal Street, Manhattan. What appeared to be a money-saving scheme, turned out to be, inch for inch, more expensive than a bridge. It took almost $50 million to build the original two-tube Holland Tunnel, about the same amount as for the eight-lane George Washington Bridge.

By 1919 traffic conditions across the Hudson were worse. World War I was over and there was a surge of economic confidence and nationalistic pride throughout the country. First the U. S. had made a fortune supplying both sides of the European conflict, then, once it had jumped into the fray, it seemed that American methods had won the war for the Allies. Now, in what was termed by President Harding "The Return to Normalcy," great American methods were easing life for the common man. Nothing, from the rankest corruption to the shock of stock scandals, could interrupt the gilded age of the '20s. Calvin Coolidge calmly directed the decade's economic expansion with blessings for private enterprise. New York had become the glamor port of the country. Manhattan was growing like an asparagus bed, with buildings like the Chrysler, Cities Service and Empire State competing for the honor of tallest.

Soon Tin Lizzies began swarming along the Hudson Valley, forming exasperating queues for the limited car space in ferries. Between 1908 and 1927, more than 15 million of Henry Ford's Model T's were sold. The original price of $950 was knocked down to $360 by the end of World War I and it later reached a low of $290, which was maintained until the Model T was discontinued in 1927. Buying on time became popular. The auto population of Bergen County, New Jersey, went up 172% between 1923 and 1928. It was obvious to all concerned that even when the interstate vehicular tunnel was completed, the internal combustion engine was going to need more facilities, and soon.

But it was an age devoted to the ideal of private enterprise, and no company could be organized which saw a profit in such an undertaking. The two states felt cautious because the Hudson River tunnel was already proving to be much more complicated and costly than anyone had expected. Had it not been for the newly created Port of New York Authority and the enthusiastic personal sponsorship of governors Al Smith of New York and George Silzer of New Jersey, the foot dragging which had surrounded the building of the Hudson River Bridge for nearly 60 years might still be going on today.

The Port Authority of New York and New Jersey, or Port Authority, as it is called by New Yorkers, has been condemned by everyone from Wall Street financiers to Henry Barnes, Traffic Commissioner under

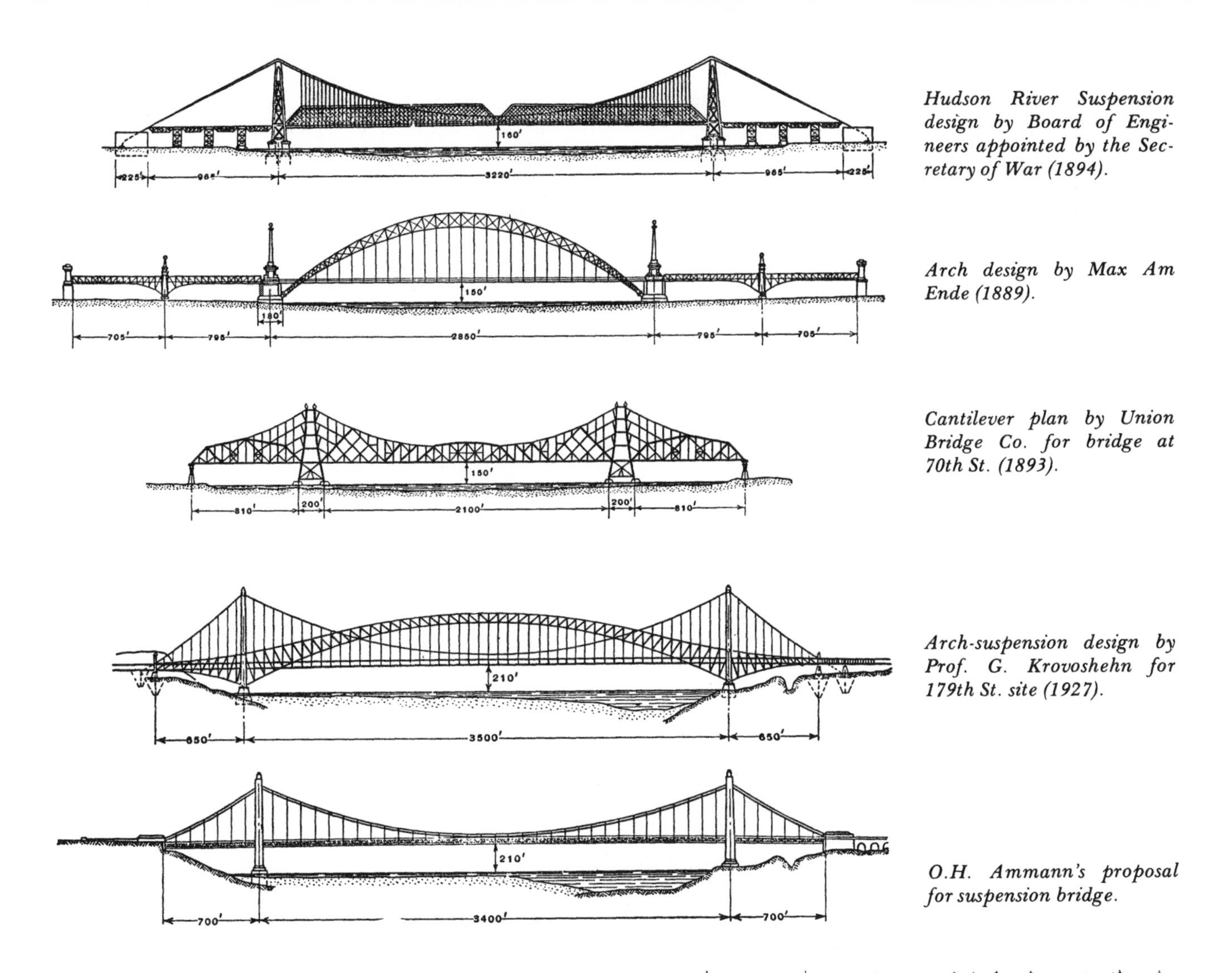

Hudson River Suspension design by Board of Engineers appointed by the Secretary of War (1894).

Arch design by Max Am Ende (1889).

Cantilever plan by Union Bridge Co. for bridge at 70th St. (1893).

Arch-suspension design by Prof. G. Krovoshehn for 179th St. site (1927).

O.H. Ammann's proposal for suspension bridge.

New York Mayor John Lindsay. "It's like an octopus with an arm reaching out in every direction, without any consideration for anyone else," said Barnes, who went on to suffer a fatal heart attack under the constant strain of unsnarling New York's traffic problems. An anonymous financier concurred with Barnes' characterization. In an article in *Business Week* in the 1960's, he criticized the Authority, whose projects are financed by tax-free bonds. He claimed the Port Authority had its fingers in so many pies that it loses sight of its bondholders in its search for bold new projects.

However, if the Port Authority is the monster many citizens believe, it is a benevolent monster. Its greatest sin seems to be its arrogance. As a public agency funded by private capital, it owes allegiance to both the public and its investors and draws criticism from both sides. Its activities can be vetoed by the legislatures of the two states but the P. A. cannot be forced to do anything. Spokesmen for the public interests deny that the Port Authority is really acting in the interest of the people and the port in general, because it will not undertake any project which might even at first lose money. The bondholders sometimes think the Port Authority gets itself involved in projects like the World Trade Center, which compete with private industry. But in its early years the Port Authority, modeled after similar institutions in London and Liverpool, seemed to be the only solution to the strife which had been all but strangling the port of New York and New Jersey.

Originally, railroads with terminals in New Jersey had their freight lightered across the Hudson to ships in New York docks. There were delays due to traffic on the river and weather conditions, naturally, but there were additional problems due to confusion, for there was no overall regulating body to promote efficiency within the port. In fact, the port was shattered into more than 200 self-governing municipalities, townships, and villages along the Hudson and New York Bay, each with its own rules and regulations. For many years the railroads had bowed to the nominal power of the politicians in these localities, for the railroads in many cases controlled them through their own strong representation in the legislatures. In 1916, however, things were changing. The enormous flow of goods eastward to supply the European war made the port of New Jersey and New York indispensable.

New Jersey townships objected to the fact that the freight rates to their state and to New York were the same, for they wanted to attract shipping to the New Jersey side. They convinced the State of New

Jersey to bring an action before the Interstate Commerce Commission which would force ships picking up freight on the New York side of the harbor to pay the lightering charges. This would establish a higher rate for the eastern side of the harbor, and give New Jersey a competitive advantage. After a thorough hearing the Interstate Commerce Commission defeated the notion of "splitting of the port", and suggested that New York and New Jersey cooperate in a full and comprehensive development of the port, situated as it was in both states.

First a bi-state commission—the New York-New Jersey Port and Harbor Development Commission—was set up, which made a study of the legal, economic and engineering problems involved. This commission recommended a port compact, or treaty, which was adopted by both states several years later on April 30, 1921. This treaty gave birth to the Port of New York Authority and the support of a comprehensive plan for developing the port as a whole. George Washington Goethals, the military engineer who had headed the Panama Canal project, was appointed to investigate the best ways of constructing terminals and connections. This was a practical approach involving a commitment to building up the port as well as managing and promoting it, a function the Port Authority still serves. It has also continued its practice of hiring the most competent men in engineering.

Within two years the Authority decided to build two bridges between Staten Island and New Jersey, the Goethals Bridge and Outerbridge Crossing. In 1923, Governor Silzer of New Jersey, in his inaugural address, reiterated New Jersey's need for a bridge to connect directly with New York City. He promptly received a design from Swiss-born engineer Othmar Ammann, who suggested that at 179th Street a bridge serving only auto traffic could be built for under $60 million. In August of the same year, Governor Silzer joined with Governor Al Smith in pledging their support for a suspension bridge across the Hudson. This bridge, they announced, would be built by the Port of New York Authority. At the time, Smith was one of the Port Authority's six commissioners.

The next year the Authority began assembling an engineering staff to work on its new marine terminals, air facilities and bridges. Among the applicants was Othmar H. Ammann, already in possession of a blueprint for the bridge he had always dreamed of. Although he was relatively unknown in political circles, he was highly thought of by most of the engineers on the East Coast. He was not only hired but was appointed the chief engineer of the Port Authority.

The man who conceived the George Washington Bridge and headed the enormous staff of engineers who made it possible had been studying maps of the Hudson ever since his university days in Switzerland. Ammann was born in Schaffhausen, a quaint Swiss town whose claim to fame was a rugged wooden suspension bridge across the Rhine, built by a village carpenter. Ammann's grandfather was a landscape painter, and from him Ammann learned the

Lindenthal's revised plan in 1920s had two decks and allowed for 16 railroad tracks plus auto lanes. It was to be at 59th St.

rudiments of art. He became intrigued by the elements of construction, especially of bridges. He thought he might become an architect. At university he found he was exceptionally gifted in mathematics and chose to study engineering but pursued no particular specialty. He became confirmed in his career of bridge engineering during a summer spent working in a bridge fabricating plant. There, Ammann did steel work and began getting a visceral feeling for the parts of a bridge. He was graduated to inspecting the parts, and finally to designing them. When he returned home he restudied the bridge at Schaffhausen.

At school that year he turned his full energies to bridge engineering. His teacher, K. E. Hilgard, had worked for the Northern Pacific Railroad in the United States. To raise enthusiasm among his students he showed photographs of America's giant bridges over the Ohio and Missisippi rivers. Hilgard's prize photo was of the Brooklyn Bridge, the greatest marvel of all. Ammann was impressed by Roebling's bridge. After "O. H.," as his American co-workers were to nickname him, founded the consulting firm of Ammann and Whitney in 1947, his office was adorned with renderings of only his own bridges, with the exception of the Brooklyn Bridge.

The greatest span over the Rhine was 750 feet. This could not compare with the great steel suspension bridge in Brooklyn, and Ammann determined to participate in some of America's important experiments in bridge building. First he finished his studies, then worked briefly in Switzerland, then in Germany. In 1902 he became restless. He mused about the enormous bridge already being planned for the Hudson. His European employers recognized a brilliant engineering talent and tried to induce him to stay by offering him a large salary increase. They also warned him that in America there was a surfeit of talented engineers. Ammann considered what they said and yet was impatient to begin a career as a major bridge builder in America. He visited his old professor and Hilgard advised him that if nothing else happened to him he would learn a great deal in America, saying, "Keep your ears and eyes open and your mouth shut." He gave Ammann some letters of recommendation. The young man arrived here in 1904.

Ammann's first employer in the United States was John Mayer, a consulting engineer. Mayer naturally had some plans for a railroad bridge in the event that he be asked to design a Hudson River bridge. Ammann studied the plans closely, noting all the problems. However, it was to be almost a quarter of a century before the project got off the ground. All during that time Ammann carefully shaped his career, learning as much as he could about bridge construction, meanwhile scrutinizing the issues surrounding the building of the Hudson River bridge.

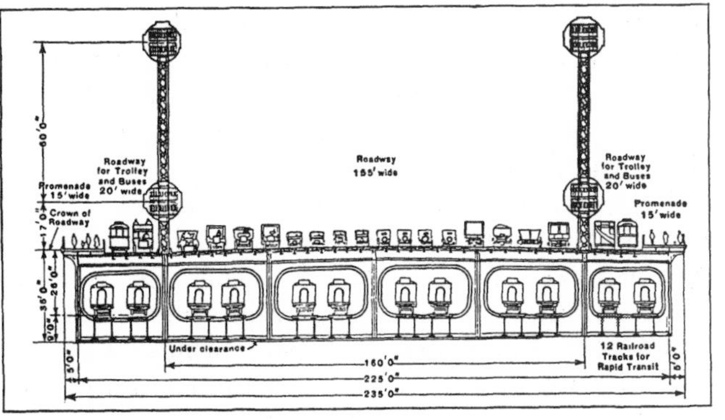

Cross section of the design Lindenthal submitted in 1923.

After helping Mayer design several railroad bridges, Ammann decided to get more experience doing detail work. To do so, he joined the Pennsylvania Steel Company, where he familiarized himself with technical detail drawings of the parts of many types of bridges. He was working there when the Quebec cantilever bridge plunged into the St. Lawrence in 1907. Ammann was consumed with curiosity about the catastrophe, for it is when such failures are minutely studied great advances in bridge technology are made. He decided he must go to the spot and study the causes, and immediately offered his services, either with or without pay, to C. C. Schneider (who had designed the Washington Bridge over the Harlem River), for the Canadian government had commissioned Schneider to investigate the disaster. Ammann headed the inquiry under Schneider's direction and for the rest of his career claimed he had gained much insight from this experience. Subsequently, he was employed by Schneider as a bridge designer for two years, but again Ammann felt the urge to move on to greater learning opportunities.

Gustav Lindenthal was preparing to build his great railroad arch over Hell Gate, and Ammann got himself recommended to Lindenthal, and wound up as second in command on the Hell Gate to the man who still was the most likely candidate to get the job of designing the Hudson River bridge. Ammann again saw plans for the Hudson River bridge and had a chance to study both the economic and structural complexities of such a span. His friendship with Lindenthal endured even though Ammann eventually got the job of designing the bridge. Ammann used Lindenthal as a special consultant on design. In fact, the two engineers arrived in the same open car at the dedication ceremonies for the George Washington Bridge.

Ammann was held in the highest regard by every engineer he ever worked for. His plans for a strictly automobile crossing over the Hudson cut down the fantastic sums estimated for a railroad bridge, and allowed for the growing New Jersey suburban traffic. In addition, much freight was beginning to be carried by trucks and because of this his bridge would ease the problem of bringing commodities directly into New York City, for New Yorkers, despite prosperity, occasionally suffered shortages. Priority on railroads at that time went to getting freight to ships bound for foreign ports.

Othmar Ammann proved to be not merely a technological perfectionist. He also had a humility and openness to suggestion which made him an able leader of other brilliant men. Unlike John Roebling and Gustav Lindenthal, who built bridges with the flair of show business pros, Ammann had an incredible faculty for drawing the best from others while underrating his own contributions. Two of his strongest competitors, Leon Moisieff (who designed the towers for the Manhattan Bridge) and Gustav Lindenthal, were called in as consultants on the design of the George Washington Bridge. Ammann was a patient man. He was dignified. He was held in the utmost respect by all who ever worked with him.

An interview with Ammann years later, after his work had included the Bayonne Bridge on Staten Island, the George Washington Bridge, Throgs Neck, Bronx-Whitestone, Triborough and Verrazano-Narrows Bridges, showed that, like his colleagues, Ammann did indeed have inordinate pride in his beautiful handiwork. He likened his favorite bridge, the George Washington, to a beautiful daughter. From his 23rd floor apartment in the Hotel Carlyle, Ammann could peer out the various windows and cast affectionate glances at his glorious steel children, inspecting them with the aid of a telescope he kept handy. Yet, when interviewed by Gay Talese about his work, Ammann smiled humbly and said it was "luck" that had made him the most respected bridge builder of all time. The year before his death, at age 84, Ammann received a National Medal of Science along with 10 other scientists. He was chosen spokesman of the group. In his usual self-negating way, he told the press that he himself was not really a scientist at all, but only made use of what the sciences had to offer. "My task," he remarked, "is to put these sciences together and build bridges out of them. Because their science is good, they stand up well."

Throughout his stint at the Port Authority, Ammann kept up his image of public servant. His low key managerial ability fit right in with the Authority's program of building bridges as efficiently as Henry Ford produced automobiles. With an organization reminiscent of an assembly line, the engineers were divided into five divisions: traffic studies, design, contracts and specifications, construction, and the planning of approaches and highway connections. In addition, the Port Authority organized a research division.

Ammann's first task was to prove that the Hudson River bridge would be profitable, for the Authority had no method of funding itself at first. The two states had authorized the Authority to issue its own bonds, which would be backed by the revenues of the facilities the Authority was building but were not guaranteed by the states. It was an era of speculation on the stock market. Fortunes were being made. Stock prices were tripling overnight. The Authority bonds were to be tax free, which gave them some appeal. But to compete with the glamour issues on the stock exchange, the Authority had to convince buyers that the enormous bridge would make money. The states backed the Authority with several hundred thousand dollars to make preliminary studies to convince prospective investors.

These studies began in July, 1925. Traffic estimates were made by 57 employees stationed on the 17 Hudson River ferries for five months. Comprehensive traffic projections predicted the number of autos that would cross the Hudson each year from 1924 to 1960. The effects of the soon-to-be-opened Holland Tunnel were taken into account. And the Port Authority received a promise from the two states that no other vehicular crossing of the river would be authorized above 60th Street, for such a crossing would be in competition with the George Washington Bridge.

The studies concluded that at an average toll rate of 50 cents per vehicle, the annual revenue would produce a minimum income of $5.25 million in 1932. That revenue in itself would be sufficient, with a substantial surplus, to meet interest and sinking-funds payments during a period of amortization of about 25 years, and the Authority would also be able to refund the initial money the states had contributed.

It is interesting to note that, despite the Depression, Americans were so enamoured of the automobile that auto population in the suburbs rose substantially through even the worst years. Traffic across the Hudson increased about 25% in 1930-31. The Holland Tunnel was constantly jammed and descriptions from the period indicate that it was being used beyond capacity less than two years after it opened.

Based on projections, it was decided to open only four lanes of the bridge to traffic for the first few years. These would be expanded to a full 8 lanes when necessary. (That was done in 1943.) The traffic forecast, which covered 37 years, showed that bridge use would continue to increase to the extent that a lower level would be needed within 40 years. Ammann decided that this second level should be designed to provide for six auto lanes or for a set of transit tracks, whichever seemed more appropriate at that future date. His plans for the lower level allowed for its construction while traffic was still flowing over the upper level. Although Ammann had assumed he would not live to see the second level built, the gaunt Swiss engineer proved to be as

COURTESY OF THE PORT AUTHORITY OF NEW YORK & NEW JERSEY

Othmar H. Ammann,in center of group of engineers, holds plans while surveying is begun on the steep New Jersey Palisades.

hardy as his handiwork. He lived to supervise the second level himself, as consulting engineer. At that time a bust of the elderly Ammann was placed in the new, avant garde 179th Street bus terminal. At the second dedication the bridge builder, embarrassed by the fuss that was being made over him, said he was only accepting the honor in the name of the engineering profession.

The first bond issue sold by First National Company was quickly bought up. It looked like a good investment. If anything, later years proved that the Port Authority traffic engineers had been overly cautious in the predictions, for from the day the George Washington Bridge opened it has carried more traffic than was originally predicted. The Port Authority has proven in the last 50 years to be one of the best bond investments in the country.

On October 21, 1927, New York and New Jersey held simultaneous groundbreaking ceremonies on both shores of the Hudson, and one in the river itself. To attract a population used to ticker tape parades for every foreign celebrity and every accomplishment since Armistice, the celebration needed a gimmick. Some inventive public relations man worked out an elaborate pageant that could have vied with the musicals of George M. Cohan. The steamer DeWitt Clinton, escorted by two destroyers and an aircraft tender with streamers flying, made

COURTESY OF THE NEW YORK MUNICIPAL REFERENCE LIBRARY

This drawing of Ammann's George Washington Bridge with single deck appeared as an illustration in the opening program.

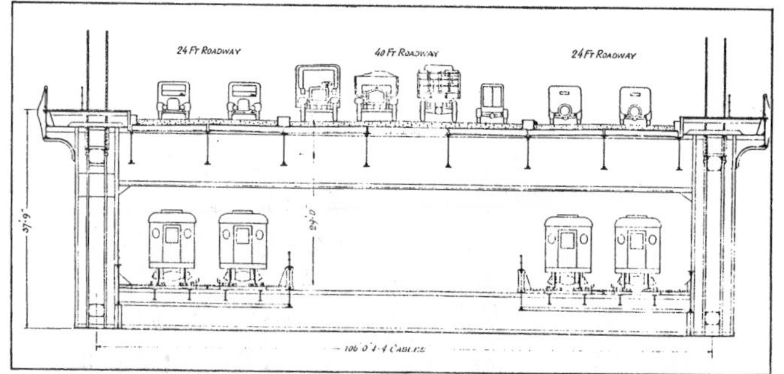

A second deck could provide rapid transit or more auto lanes.

its way up the river carrying 1,000 public officials from New Jersey and New York, representing various civic organizations which had fought so long for the bridge. Lunchtime crowds lined both sides of the river. At 42nd Street two seaplanes from the Naval Air Station at Hampton Roads joined the coterie and performed aerobatics above the flotilla. After their performance was finished, they moved off and circled in wide sweeps high above while a second contingent of planes, this time from the Army, flew closer to the ships in formation and demonstrated aerial combat maneuvers. Finally, the DeWitt Clinton anchored midstream at the site of the future bridge and at the New Jersey bridge site, high on a cliff of the Palisades draped with a huge American flag, there was a "daylight fireworks display," featuring bombs shot into the air, breaking and showering clusters of American flags, vari-colored smoke and dummy parachute jumpers. The thrill of postwar patriotism infused the entire proceedings.

Bands blared on both sides of the river. Officials on each shore put their shovels into the soil. The shovels were later carried to the DeWitt Clinton and presented to Governors Al Smith and Arch Moore and to the new chairman of the Port Authority, George S. Silzer. The radio broadcast of the speeches from the steamer was held up when a passing ship cut the cables, but after a 45-minute delay Al Smith, in a playful mood, noted to his Republican colleague that it would soon be easier for New York's Democrats to slip into New Jersey. The Port Authority's engineer and Chairman Silzer pledged that the new monumental bridge would not ruin the magnificent natural scenery. The speakers all expressed the hope that the bridge would be open by the intended date in 1932. In fact, the building of the George Washington Bridge was so efficient that it opened eight months ahead of schedule, in October 1931.

Ammann's leadership of the operation proved effective. Although the engineer maintained that anyone who took complete credit for building a bridge was "an egotist", the George Washington Bridge is his visual conception. The vast structure manages to show its strength without looking cumbersome, gross or unwieldy. "An engineer is justified in making a more expensive design for beauty's sake alone," said Ammann, who suggested that his original design could stand by itself aesthetically without being clothed in granite-faced concrete, as originally planned. By 1931 the Depression was on and Ammann noted that this addition would have caused the bridge to exceed its original $60 million estimate, while the Port Authority had managed to keep construction costs considerably under that.

The George Washington Bridge is unquestionably the strongest suspension bridge in the world and has maintained this record for over 40 years, although it has been demoted to the rank of fourth longest suspension bridge. How did Ammann manage to let it retain its powerful grace despite its weight requirements? "Simplicity," said the Swiss engineer. "Simplicity," he repeated in an article for the Smithsonian Institution during the building of the bridge. And so, without fanfare, marble arches or decorative moldings, Ammann let the bridge's structure speak for itself.

The George Washington Bridge contains 113,000 tons of fabricated steel, 28,000 tons of cable wire and 200,000 cubic yards of masonry. Each of its steel towers incorporates two arches, one above and one below the roadways. The arch above is as high as a 17-story building. The total length of steel wire used in its four-foot diameter cables was 106,000 miles, or almost half the distance to the moon. Yet no one gaped, as they had for the Brooklyn Bridge, or feared for its safety. Few were impressed; it was just another step in the advance of progress; one of the spiraling statistics that seemed to accompany all the new buildings, automobiles and airplanes. New Yorkers were too blase to be impressed by the fact that the New York tower stood on the greatest mass man had ever built on so limited a space, and that the new bridge was "making a pigmy" of the Brooklyn Bridge, as one magazine put it.

But engineers were taking note. Around the country, municipalities were reevaluating their decisions on bridge building. San Francisco started planning the Golden Gate, for which they called in Othmar Ammann as consulting engineer. This beautiful West Coast bridge, which has attained fame beyond New York's own George Washington and superseded it in length, could not have been built without findings Ammann arrived at in designing and building the George Washington Bridge.

The last previous comprehensive investigation of the maximum length of a suspension bridge had been done by the Army Corps of Engineers in 1890 regarding the proposed Hudson River Bridge. Ammann's investigation went far beyond that one. The George Washington Bridge answered questions about the feasibility and economy of a span of such unprecedented length. It demonstrated that a modern suspension bridge of even 10,000 feet span could be built with the normal factor of safety. "We may now refute the conception held even by engineers that the length of span is the major economic factor in the construction of a large bridge, in that it is supposed to influence the cost about in proportion to the square of the span. Traffic capacity and the cost of foundations and approaches are apt to be far more important economic factors than mere length of span," Ammann said. Ammann, in fact, contended that a suspension bridge 10 miles long could be constructed which would physically support itself; however, it could not possibly carry enough traffic to be economically self-supporting.

To get absolutely economical work from contractors, Ammann's preliminary studies made extensive use of his research department. This department was not limited to abstract mathematical calculations and drawings. In addition there were physical facilities equipped with wind tunnels where scale models were submitted to simulated weather variations to test stresses and strains of various designs. When several cable arrangements proved suitable for the bridge, the contractors were asked to submit competitive bids on alternate designs. Possibilities included cables placed side by side or one above the other, composed of eye bars or of wire, and strung simultaneously or successively. This gave bidders wide latitude in applying their ingenuity and experience toward developing the most economical methods of fabricating and construction.

One engineering discovery made in the course of design research contributed to the good looks of the George Washington Bridge. "Based on comparison and on the theories we find in textbooks and other treatises on suspension bridges, we should expect the stiffening system of the George Washington Bridge to weigh from 13,000 to 14,000 pounds per foot and be 11 stories high," Ammann told a group of engineers several months after the bridge opened. "Actually, it weighs only 1100 pounds per foot in the initial stage with only one deck for highway traffic and will weigh 2350 in the final stage with two decks."

When Roebling built the 800-foot suspension bridge at Niagara Falls, he realized that as a train advanced over it there would be a local sagging of the cables so that a wave of depression would mark the progress of the train. To overcome this hazard he incorporated a deep truss along the roadway. Ammann found that the Hudson River bridge would

COURTESY OF THE PORT AUTHORITY OF NEW YORK & NEW JERSEY

Construction site in winter shows roadbed being hung from suspenders. Prefabricated parts are raised from anchored barges.

be so heavy that an extraordinarily shallow truss would suffice to prevent vertical distortion of the roadway. The shallow truss on the roadway protects against high velocity winds, but the enormous inertia of the bridge itself provides the principal resistance. This natural rigidity is aided by the short side spans, where the sharp angle of the cables enables them to act as rigid back-stays. Research has shown that this inertia is so great that if the bridge were struck by a furious gust, such as might come in a thunderstorm, the wind's force would be spent before the bridge moved appreciably in response to it. In the improbable event of a continuous wind of hurricane force, the center of the bridge would move no more than 22.18 inches from normal.

The truss between the original deck and the lower deck (completed in 1963 and playfully referred to as Martha) has diminished the graceful quality of the roadway, but the handsomeness of the bridge is still apparent. Instead of being surrounded by a clutter of wharves and warehouses, like most of the East River bridges, the George Washington soars between grandiose cliffs and parks. Rather than building a massive cement anchorage above the Palisades, Ammann chose to anchor the wires in the cliffs themselves. He had the workers dig a deep inclined tunnel into the rock. In this tunnel, big enough to accommodate four trolley tracks, eye bar chains were placed, and the wire—26,474 double galvanized wires for each cable—was attached to the chains and the entire tunnel was filled with cement. On the New York side a huge anchorage was built in Fort Washington Park, which serves as a roadway arch. The trap rock dug from the New Jersey cliffs was used to make cement for the foundations of the two towers. The New Jersey tower was

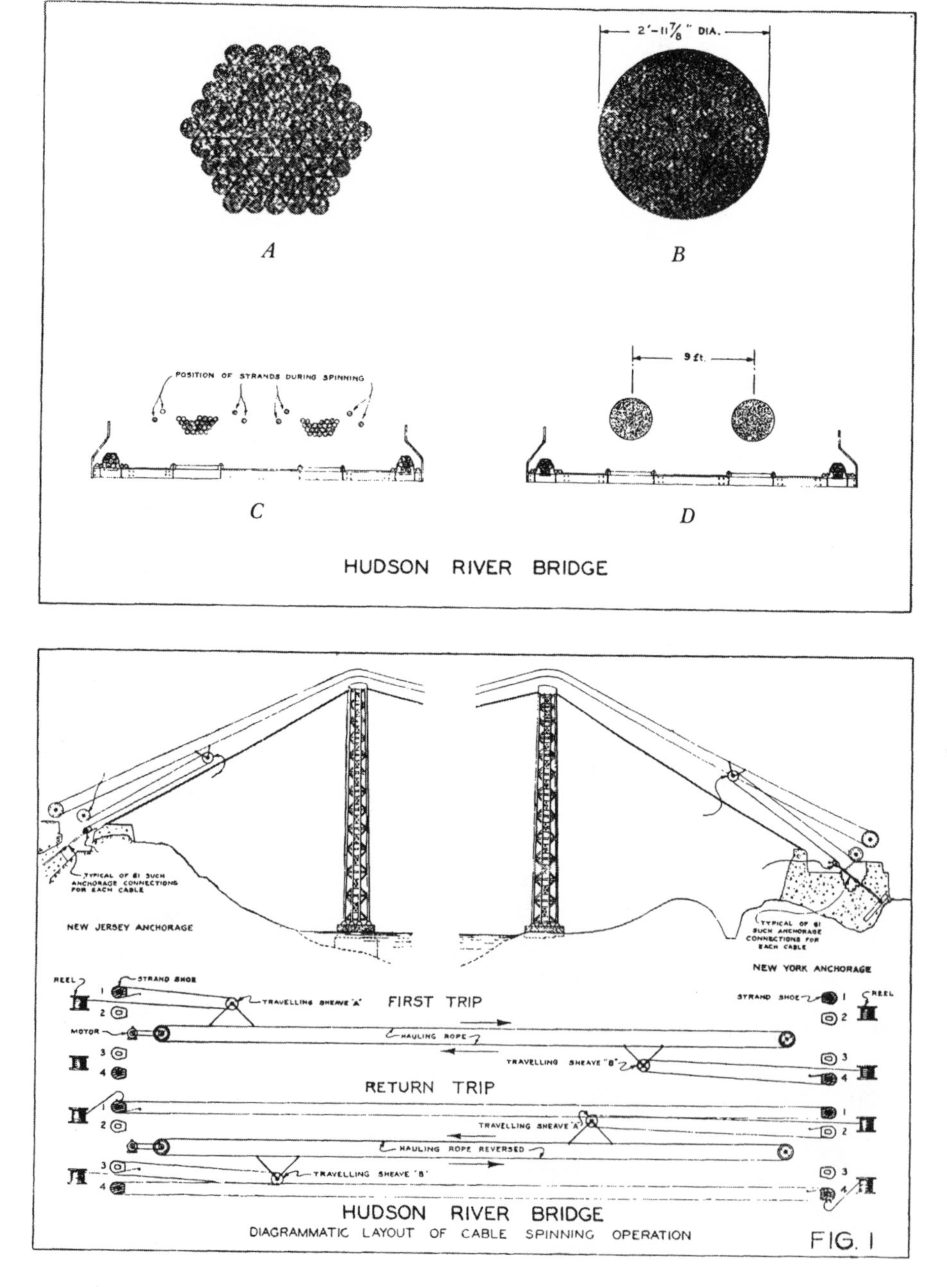

Figure A: Cross section of cable before compacting. Has 61 strands of 434 wires.
Figure B: Cable after compacting has a diameter of 2 feet 11-7/8 inches.
Figure C: Diagram of partly completed cables shows footbridge and position of strands during spinning.
Figure D: Compacted cables are placed 9 feet apart.

Cable spinning operation is shown in this diagram.

located 76 feet into the river while the New York tower was built on the land, to avoid the deep fissure in the river.

For the first time in New York, an entirely professional bridge crew was employed to build the structure. Like the engineers who fought to design the great bridge and the contractors who tried to outbid one another, these men had great competitive spirit. They were divided into three separate crews, one for each tower and one for the roadway. The rivalry was so intense that although the New Jersey tower was started on May 16, 1928 and the New York tower was begun on July 13, the New York gang announced it would be the first to complete construction anyway. It did. Of course, the New York tower was more accessible, as it was on land, which made transfer of materials from the ground to the tower easier than on the New Jersey tower, where material had to be ferried out.

There were 12 deaths altogether during the construction of the George Washington Bridge. While the foundation for the New Jersey tower was being dug, a cofferdam was flooded and three men were drowned. A fourth fatality resulted from a premature explosion of powder while blasting the rock from the New Jersey anchorage. The rest were due to the carelessness on the part of the workers, according to Port Authority records.

The Port Authority did a professional job of publicizing the entire venture, and even had a movie made of the highlights of the building of the big bridge. A still photographer preserved all the history-making operations of the bridge's construction. While the workmen were busily preparing the 635-foot towers, which dwarfed all existing bridge towers, and paving the spaghetti-like complex of exits and entrances, the public was consumed by two issues concerning the new bridge: What are we going to call it? Are we going to encase those giant steel towers in a sheath of concrete?

The name George Washington Memorial Bridge had been proposed by a member of the Port Authority in the early stages of construction because the bridge site is historically famous for Fort Washington and Fort Lee, the forts on either side of the Hudson where General George Washington had made a stand against the British during the Revolution. The word memorial was used to commemorate the men who had died in the Great War. But many citizens were piqued. After all, there was already a nearby Washington Bridge spanning the Harlem River, and the name might cause confusion. The *New York Evening Post* crusaded for naming it "The Palisades Bridge," while others said no matter what name was given to it, it would probably still be called the "Hudson River Bridge."

When the Port Authority, which tries in such matters to be sensitive to public opinion, heard the

COURTESY OF THE NEW YORK MUNICIPAL REFERENCE LIBRARY

Drawing made for prospective investors shows Cass Gilbert's granite tower design. It was abandoned during the Depression.

outcry against the name George Washington Memorial Bridge, it opened the matter to public ballot, and all interested citizens were invited to send endorsements or new suggestions. The result of this poll was that the original name, George Washington, received the most votes. Palisades Bridge came in second, then Hudson River Bridge. A few people petitioned to have the bridge named Alexander Hamilton, General Pershing or Verrazano. It happened that most of the votes for George Washington came from school age children who affixed their signatures to petitions. So, with the juvenile vote on its side, the Port Authority decided to keep the name George Washington to avoid "insulting the memory of our first President and encouraging the Reds," as then Congressman LaGuardia said. The Port Authority did drop the word Memorial from the name.

The second controversy got all New York embroiled in a dispute about aesthetics. Ammann had designed the tall towers to alone provide enough support for the two roadways which the bridge was to carry. The stone dressings were to be added in a separate operation. The Port Authority had hired as consulting architect for the "memorial bridge" Cass Gilbert, the eminent architect who had designed the Woolworth Building. In keeping with Ammann's theories about simplicity, Gilbert designed a granite faced cement dressing for the towers which would give them a more monolithic and "lasting" quality. He also proposed that there be restaurants and observation towers atop the piers.

However, as the towers were completed, those who came to the bridge site grew more and more attracted to the towers' "deceptive lacelike frailty," as the *Nation* put it.

"It is hardly possible to enter into a discussion of the reasons and considerations which eventually led to the decision to build the towers initially as a steel frame designed to carry the entire load and to leave it for future consideration as to whether the frame is to be encased in or surrounded by stonefaced concrete," Ammann said in 1932. By then, however, much had changed. The nation had been on a mad buying spree for a decade. Now the Empire State Building was a white elephant and people stood on breadlines. The Port Authority had just bought the Holland Tunnel from the Interstate Bridge Commission for $48.5 million—about the cost of construction—and was in no mood to spend additional funds for decorative purposes. Perhaps the Roosevelt Administration could have been convinced to pay for the stone work as a public works project. But in any case, there was no lack of aesthetic quality in the gigantic million-riveted lattice-work towers. Most newspapers and journals admitted that the naked towers and far flung roadway were indeed a magnificent addition to the river and no stonework was necessary. Leon Moisieff, a design consultant on the George Washington Bridge and the Manhattan Bridge towers, noted that the time was gone when people thought only stone could give a monumental appearance. Steel, too, could now serve this purpose.

Anyway, the reasoning went, it is not impossible that at some time in the future these towers might be clothed. The *Times* noted, "Ours is a utilitarian age, of course, and one afflicted at the moment with a disease called a Depression. But it is also an age with a powerful urge toward experiment. Economic power is not wholly responsible for the change in the bridge piers. The economic arguments are strong, but something else: the effect of the steel beams on the landscape. The notion has got considerably diffused that what has already been achieved in carrying out the monumental design in steel provides an eyeful that could hardly be bettered by trying to make steel towers look like stone piers—even stone piers designed by the architect of the Woolworth Building and so much else that is fine and distinguished. Of course, we don't intend to tie up the future to current notions of aesthetics and the stone dressing can be added at any time in the future."

Unlike the towers on most modern suspension bridges, the George Washington Bridge towers are not flexible, with columns lined up on one plane. Instead, the towers are composed of rigidly connected three-dimensional frames with arched portal bracing.

COURTESY OF THE PORT AUTHORITY OF NEW YORK & NEW JERSEY

Franklin D. Roosevelt was governor of New York when he spoke at Dedication Day Ceremonies in Oct. 1931. Eleanor attended too.

Ammann himself was not so bold as to make any direct appeal for leaving the towers bare. Calling on the Port Authority to pay heed to public opinion, Ammann outlined his own concept of aesthetics.

"The writer (Ammann), who has conceived and is primarily responsible for the type and general form of the design, considers the steel towers as they stand to represent as good a design as may be produced by a slender steel bent, and that this lends the entire structure a much more satisfactory appearance than he (and perhaps anyone connected with the design) had anticipated. Nevertheless, he believes that the appearance of the towers would be materially enhanced by an encasement with an architectural treatment, such as that developed by the architect, Mr. Cass Gilbert.

"The writer is not impressed by the criticism, based solely on theoretical and utilitarian grounds, that the encasement would constitute a camouflage which would hide the true structure and its function. The covering of the steel frames does not alter or deny their purpose any more than the exterior walls and architectural trimmings destroy the function of the hidden steel skeleton of a modern skyscraper, except to the uninitiated.

"Camouflage in this sense would condemn many of the creations in private and public life. It is an essential manifestation of civilization and is not incompatible with sincerity and honesty of endeavor, because an essental part of human effort is to create an aesthetic atmosphere, the value of which cannot be expressed in economic terms. This is evidenced in the craving for beautiful homes and public institutions which yields only to the limits of available means. Why should not a supreme effort be made in that respect in engineering structures, especially those which are viewed daily by thousands or millions of people?"

But despite this philosophical and fulsome defense of a monumental encasement by Gilbert, Ammann admitted that he would be "satisfied" if the structure was left as it was; that his efforts were "not without fruit," that the steel towers as they stood had a "good appearance, a neat appearance," owing to "sturdy proportions and well-balanced distribution of steel in the columns and bracing."

So, at the dedication of the George Washington Bridge on October 25, 1931, members of the President's cabinet, Governors Franklin Delano Roosevelt of New York and Harry Moore of New Jersey, as well as 5,000 guests of the Port Authority sat in the shadow of two openwork steel towers. It was a breezy Saturday afternoon, and bleachers filled to capacity with invited guests ran the length of the bridge on the side of the roadway. Crowds filled the New York and Fort Lee Plazas, eager to get a glimpse of presidential nominee Roosevelt, and delighted to have an amusement that cost nothing.

According to the most famous story dealing with the dedication ceremonies, when the officials were gathered at a luncheon at the 102nd Engineers Armory at Fort Washington Park before the ceremonies, a skeptical National Guard Colonel asked Ammann if the bridge would support the military parade scheduled to cross it. From time immemorial, troops had marched at "route step" over bridges for fear that rhythmic footsteps would set up harmonic vibrations and collapse the bridge. The Swiss engineer wryly assured the officer that the bridge would stand up if an army of elephants marched over it. So, defying the age old tradition of breaking step when crossing a bridge, columns of soldiers, sailors, marines and coast guardsmen came swinging down the roadway from the Manhattan Plaza. The guests at the center of the bridge are reported to have felt the gigantic span vibrate as if shaken by earth tremors, but they were only amused by it. There was no panic.

The Secretary of the Navy and other cabinet officials were in attendance, as was Al Smith, former Democratic presidential nominee, whose participation in the building of the bridge was praised. The loudest and most spontaneous applause, however, was for Othmar Ammann and Gustav Lindenthal, who wheeled across the bridge in an open car.

The bridge had been finished eight months ahead of schedule and for less than the $60 million estimated by the engineers. No one had anything but praise for the Port Authority and its staff. More than 30,000 spectators watched the governors and other officials as their speeches were amplified to the New Jersey and New York plazas and carried over radio. Thirty-five planes flew in formation overhead, and

Bleachers were set up on the bridge for the Dedication and 5,000 guests were invited.

COURTESY OF THE PORT AUTHORITY OF NEW YORK & NEW JERSEY

COURTESY OF THE PORT AUTHORITY OF NEW YORK & NEW JERSEY

The first structural steel section for the lower deck of the George Washington Bridge is raised in 1960.

one of them delighted the crowd by diving under the span. Debonair New York City Mayor James J. Walker, then under investigation for corruption by the Seabury Commission was conspicuously absent. He sent regrets, but said he had a previous engagement at a Yankee Stadium football game.

After the speeches, the ribbons were cut on either side of the bridge and it was opened for the afternoon to pedestrians. The first across were two Bronx boys on roller skates. Actually, these were not the first "civilian" pedestrians to cross. In June, the Port Authority had granted this honor to 40 bankers and financiers who had been instrumental in financing the bridge. Despite the rain, these men had proudly trooped over the bridge and had been escorted to the top of the towers in elevators.

After strolling across the bridge on dedication day, people attended parties held on both ends of the bridge. From midnight to five the next morning the bridge was closed so debris could be removed to clear the way for auto traffic. The next day 56,312 cars crossed the bridge.

One ironic aspect of the toll rate on the George Washington Bridge was that it cost more to walk the ¾-mile span than to take the bus across. While the bus ride from Manhattan to Fort Lee cost a nickel in 1931, the pedestrian toll was a dime.

In the next 40 years the George Washington Bridge became the almost-villain of a classic children's book called *The Little Red Lighthouse and the Big Gray Bridge,* by Hildegarde Swift. The story's hero, an old-fashioned lighthouse in Riverside Park, feels its usefulness is usurped by the fog lights on the towers of the George Washington Bridge. It is consoled, and again made cheerful, when told that the bridge's lights are for airplanes, not ships. This lighthouse actually was considered obsolete and was ordered to be removed, but was saved by strong public opinion and the leadership of Robert Moses, Commissioner of Parks, in whose jurisdiction the lighthouse stood.

The George Washington Bridge also inspired a composition for band by contemporary composer William Schuman, who was president of the Lincoln Center for the Performing Arts in New York. Mr. Schuman, who composed the piece in 1950, wrote: "Ever since my student days when I watched the progress of its construction, this bridge has had for me an almost human personality, and this personality is astonishingly varied, assuming different moods depending on the time of day or night, the weather, the traffic and, of course, my own mood as I pass by. I have walked across it late at night when it was shrouded in fog, and during the brilliant hours of midday."

The New Jersey arch of the George Washington is rigged on holidays with the world's largest free flying flag. The stripes are five feet wide and the stars are three feet in diameter. The entire flag is 60 by 90 feet and weighs 475 pounds.

In August, 1962, the second deck was finished. The next year a new bus terminal designed by Pier Luigi Nervi, an Italian engineer, was erected on the Manhattan side at 178th Street. The reinforced concrete wing-roofed building, one of the most unusual in New York, has a capacity for 200 buses and over 10,000 people in peak rush hours. It replaced the scattered individual terminals in the congested 166th-167th Street area of Washington Heights. Nervi's open roof design permits ventilation of bus fumes. The interior of the lower level, in contrast to the stark modern exterior perpetuates the peculiar habit of the Port Authority of decorating interiors with bathroom-type tile. This building also houses the bronze bust of Othmar Ammann, dedicated the day the second deck of the George Washington Bridge was opened.

The second dedication ceremonies were led by Governors Nelson Rockefeller of New York and Richard J. Hughes of New Jersey. They crossed from opposite ends of the lower deck in restored 1931 touring cars, replicating the atmosphere of the earlier opening of the George Washington Bridge.

The method of construction of this lower level had followed Ammann's original design. Seventy-six structural steel sections for the lower level were raised from the shores or from barges below without interrupting traffic on the upper span. The only problem that presented itself was the need for a slight adjustment of one of the rollers on the New York towers.

The most expensive aspect of the new addition, which brought the total Port Authority investment in the bridge to $219 million from the original $60 million for the 4 lane single deck bridge, was the approaches to the second level. Ammann noted that if he had foreseen that the second level would definitely be for auto crossings he would have left more room for such highways. The new blasting through the Palisades required plenty of time and money. The second level and approaches took from 1958 to 1962 to build, more time than the original bridge.

The *New York Times,* in reporting the second dedication, expressed the feeling of many New Yorkers about the bridge and about Othmar Ammann, who was now working on the world's biggest suspension bridge further down the river.

"He (Ammann) was the dreamer, he was the artist, he was the solid and reliable planner who made this beautiful structure possible and durable.

"An artist indeed he is. Someone has said it is impossible to make an ugly suspension bridge. The towers and curves of the supporting cables solve a problem of aesthetics as they do in engineering. But the George Washington is something special. Consider this bridge viewed from the north,, with the river shining below it in sunlight, moonlight or starlight with the man-made towers of Manhattan on one side and the Palisades on the other. This bridge is not steel alone; it is not merely a machine or an engineering device built for the purpose of crossing running water, feeding traffic into bottlenecks. No—it is as much a picture as any that hangs in any museum. There it stands, framed for uncounted years by land and sky. It is convenient to drive over. It is also inspiring to look at, in the realization that when man builds strength he may also build beauty."

COURTESY OF THE PORT AUTHORITY OF NEW YORK & NEW JERSEY

World's largest free flying flag hangs from tower at dusk.

COURTESY OF THE PORT AUTHORITY OF NEW YORK & NEW JERSEY

The elegant lines of the longest steel arch bridge in the world. The Bayonne Bridge was designed by Othmar Ammann.

STATEN ISLAND'S BRIDGES

OF HILLS AND KILLS

The Arthur Kill and Kill Van Kull, two connected waterways which separate Staten Island from New Jersey, have been called an "Othmar Ammann lover's delight." Unfortunately, this description is misleading. Of the four bridges which cross the oily, industrial waterway, only one, the Bayonne Bridge, was designed and overseen by Ammann. This bridge is the world's longest steel arch. Its harmonious patterns of interlocking triangles and the streamlined arch reveal Ammann's sensibilities. The remaining three bridges are commonplace. They include a railroad lift bridge owned by the Baltimore and Ohio Railroad, and two narrow cantilever spans. The Goethals Bridge between Elizabeth (New Jersey) and Howland Hook (Staten Island), and the Outerbridge Crossing, between Tottenville (Staten Island) and Perth Amboy (New Jersey) are run by the Port Authority and were designed before Ammann joined the Authority's staff in 1924. Plans for these two cantilevers were completed by John Alexander Waddell, whose firm, Waddell and Hardesty, served as consultant to the Port Authority before the organization built up its own staff.

Those visitors to New York who do not take New York City geography for granted, immediately notice the incongruity of Staten Island being part of the city, or even part of New York State. Geographically and geologically, the 61-square mile island is more akin to New Jersey than New York. The Kills between the meadowlands of Staten Island and the mainland are considerably narrower than the distance which separates the island from Manhattan or Brooklyn. Also much of the underlying rock is a continuation of the Palisades of New Jersey.

There are two explanations for how the fertile island wound up as part of New York State. One widely accepted story is that a wily Tory, Christopher Billope, a Staten Island landowner who favored its joining New York, commandeered a small sailing ship and raced around Staten Island within the 24-hour time limit supposedly prescribed by the Duke of York to make any land mass a part of New York City. Historians call this story a 19th century myth. Actually, there was a lengthy dispute in colonial times about whether or not Staten Island did belong to New York.

When it first apportioned territory around New Amsterdam, the Dutch West India Company awarded Staten Island to a man named Michael Pauw. Pauw also was allocated the land in what is now Bayonne, Hoboken and Jersey City. However, Pauw never established a permanent settlement on the island, and his patroon on the Jersey peninsula was overrun by Indians. These lands changed hands several times. When the British took over Staten Island in 1661 only 14 white families lived there. The British renamed the island Richmond and redistributed the land in such a way that its jurisidiction became cloudy.

The Duke of York, absentee governor of the crown's new colony, glanced at a map of the New World in the comfort of his English manor and bestowed the land west of the Hudson to Lords Carteret and Berkeley. Concomitantly, he sent Richard Nicoll to govern New York. Nicoll, unaware of the Duke's gift, decided the boundary line between New York and New Jersey should be the Arthur Kill and Kill van Kull, streams which may not have been marked on the Duke's map. Thus, two claims to the island of Richmond resulted. The governors of New York were more powerful than their New Jersey counterparts, who were involved in boundary disputes with Pennsylvania as well, so Richmond became part of New York.

Although annexed to New York State, Richmond ignored neighboring Manhattan and Long Island, displaying a natural affinity for New Jersey. The first island ferry service started in 1671, when sailboat ferries ran from Elizabethtown, New Jersey to Howland Hook, through the narrowest portion of the Kills.

Richmond, as an agricultural island, was really self-contained during colonial times. This isolationism was reinforced during the Revolution when the island became headquarters for General Howe's Army and the British Atlantic fleet. Permits were needed for islanders to cross to New Jersey or to New York. Staten Island's first bridge to the mainland was built by the British during this occupation. The pontoon bridge crossed from Elizabeth to Howland Hook. Although General Washington controlled New Jersey, Elizabeth was difficult to defend, as the Redcoats could slip across the pontoons into New Jersey. To cross, the soldiers often had to duck bullets and sometimes even crawl. So it became known as "belly bridge." Civilians using this bridge were suspected of spying. After the war the bridge was removed.

Staten Island has always seemed remote to other New Yorkers. Until after 1800 there was no regular ferry service from Staten Island to Manhattan. Then Cornelius Vanderbilt, a Staten Island resident, started his career as a transportation magnate by running a piroge (an early type of ferry boat) from Staten Island to New York in 1810.

The first influential person to promote the island was Daniel Tompkins, vice-president of the United States under James Madison in 1817. He invited guests to the lovely island to escape New York epidemics and heat, and enjoy the island's good air and scenery. The islanders tended to call these summer people "foreigners." The "foreigners" spent the summers at New Brighton, which was built to rival in splendor its English namesake. New Brighton's posh Greek revival Pavillon Hotel was accorded the accolade of the most elegant hotel in the New World.

Although Staten Island was becoming established as a fashionable resort area, it needed transportation service to achieve a stable growth rate so Tompkins laid out new roads and began to advertise the sparsely settled island as a stage route to Philadelphia. As the route was shorter than the one from Northern New Jersey, travelers were urged to ferry to the island in the evening, stay overnight at an inn, then catch the morning stage. The ferry was the famous steamboat *Nautilus,* which plied between Tompkinsville and Whitehall Street starting in 1817. The *Nautilus* also became New York's first tugboat.

This ferry system was suitable for the wealthy sophisticates who began building vacation homes on the island's northern shore. But the backbone of Staten Island's agricultural population traded for the most part with New Jersey, by means of the three ferries which had sprung up along the Kills. They ran regularly between Tottenville and Perth Amboy, Elizabethtown and Howland Hook, and Port Richmond and Bayonne.

The first vigorous campaign for a bridge came from new residents and summer people in 1850. They considered five daily summer ferry crossings from Staten Island to New York insufficient, and believed a link to Bayonne would enable them to use the more regular Bayonne-New York ferry. Such projects were opposed by year-round residents, who were satisfied with their island homes and desired separation from other communities. They said the bridge advocates were not interested in bettering the island, but in more convenient transportation to their businesses in New York City. But in 1870, the New York State Legislature passed a law permitting a bridge from Ellis Island over the shoals to Robbins Reef and from there to the heights of New Brighton by means of a swing span. The bridge required no remarkable engineering innovations, as the shoals were only about five feet deep. At the Ellis Island end of the four-mile long bridge, a ferry terminal with five slips would carry passengers and traffic to downtown New York. It was expected that later a span could be built from Ellis Island to a New York terminal. A charter was granted to the Staten Island Bridge Company for the Robbins Reef span. As the United States was in the midst of one of its cyclical depressions, however, funds were lacking and the project was abandoned.

It was then that Erastus Wiman, a crafty Canadian-born financier and real estate speculator, saw tremendous development possibilities for the isolated island. He established the Staten Island Rapid Transit Railroad, which ran to Vanderbilt's Landing and oversaw construction of the original St. George ferry terminal in 1886. This famous structure has been rebuilt twice, the second time after a disastrous fire in 1946. The 1951 replacement is familiar to many New Yorkers and visitors from all over the world who ride on the Staten Island ferry. St. George, incidentally, was named after a Manhattan business man, not a saint. In order for Wiman to get financier George Law to back the ferry terminal and develop the surrounding community, Wiman promised to cannonize him. Hence the appellation St. George.

Wiman believed a direct connection from St. George to New Jersey was imperative and extended his railroad along the northern shore of the island to Howland Hook. The next problem was to get a bridge built. Wiman approached the Baltimore & Ohio Railroad which was interested in an outlet to New York Harbor. In a contract the railroad agreed to finance the bridge in return for the use of Wiman's railroad facilities on the island. Wiman, in addition, had to increase the ferry service from St. George to New York. The B. & O. petitioned for Federal permission to build the bridge, as the Kill is an interstate waterway. The design was for a swing span with a center pier. The Senate agreed to pass the bill, and it was signed by the President in 1886.

The railroad swing bridge was 800 feet long. Its draw was 500 feet and there were two 150-foot side spans. It was only 32 feet above low water. The cost

THE ARTHUR KILL BRIDGE

COURTESY OF THE STATEN ISLAND HISTORICAL SOCIETY

The first span over the Kills was this swing bridge. It carried its first train in 1890 and was replaced in 1959.

of the project was half a million dollars.

This was Staten Island's first connection with the surrounding communities, and it was quite an event for the sleepy island. The opening ceremonies were set for New Years Day, 1890. Two cars were attached to a locomotive, decorated with flags and bunting and filled with invited guests, officials and newspapermen. The train left St. George at 11 a.m. Among the occupants of the two cars were Erastus Wiman, President J. Frank Emmon of the B. & O., and Chief Engineer Charles Ackenheit, builder of the bridge. Mr. Wiman piloted the train over the bridge while two carloads of passengers cheered and steam whistles tooted. Bridge and railroad workers yelled and adventurous boys climbed on the structure. A stop was made in the middle of the bridge to allow the passengers to inspect it. The traditional oratory at such events was made where the railroad connected with a B. & O. branch line in New Jersey. Although this bridge has never had any regular passenger service, during World War I and II it carried to Staten Island piers thousands of troops bound for Europe.

More recently, after dredging operations on the Arthur Kill, large ocean going vessels and numerous smaller ones passing through the strait resulted in numerous bridge openings each day. A 1952 record indicated that 13,446 vessels required the bridge to open during that year. The bridge's machinery was steam powered, and in foggy weather a steam whistle signaled whether the span was open or closed — two blasts meant the bridge was closed and four that it was opened. In November, 1952, heavy winds caused the span to spin around and the two engines could not close the bridge for some time. Another time an ocean-going steamer with a jammed steering gear crashed into a section of the Staten Island trestle and carried it forty feet downstream on its bow.

As the bridge pier was an obstruction to navigation, the government decided during the 1950s that the old swing span must be replaced by a modern lift span. The government argued that "if the old bridge should become disabled, more than 100 Staten Island industries would be cut off from rail communication with the rest of the nation." The Arthur Kill was deepened further in the meanwhile, and contracts were drawn up for a new lift bridge 150 feet east of the original span. The Federal Government paid 83% of the cost of the new bridge. The distance between the towers of the new bridge is 558 feet, which made it the world's longest vertical lift bridge at that time. The clearance of the new bridge is 135 feet. The first train to cross consisted, not of passenger cars, but of a blue diesel engine pulling 65 empty coal cars from Staten Island to New Jersey on August 25, 1959. The new bridge enables the B. & O. to use 70 ton cars instead of 50 ton cars. The old bridge was demolished and its granite piers were blasted away.

The original Staten Island Railroad Bridge remained the island's only physical link with the mainland for over 30 years. Although Staten Island voted to join with the other boroughs to form Greater New York in 1898, residents were not really keen on building a bridge to New York. Ferry service was improved, and only industrialists and merchants vigorously backed a bridge project.

For New York City after consolidation, the building of a connection to Staten Island was considered a very expensive undertaking. The city was already engaged in building bridges to Brooklyn, Queens and the Bronx and was contemplating a Hudson span. It had no desire to add to its obligations by building a bridge to Staten Island. Staten Island was termed the "Cinderella Borough" by the *Times*. But by the end of World War I the industrial age was hitting even the rural marshlands of Richmond. Along the Kills factories and refineries were popping up. The three ferries to New Jersey were being taxed beyond capacity with freight, commuters and automobiles.

It was during the post-World War I building boom that serious pressure was exerted for bridges between Staten Island and either New York or New Jersey. Bills to secure such crossings had been introduced since 1868 in both New York and New Jersey. In annual speeches from 1910 to 1919, Mayor Mravlag of Elizabeth had pleaded for such a bridge. But no action was taken until the Port of New York Authority was formed, because there was no instrument by which the states could get together and build a joint project.

In 1917, New York mayor John Hylan began to advocate developing Staten Island. After all, Hylan noted, Richmond composed 1/6 the land mass of the entire city, yet it had only 1/60 the population. Hylan got some money appropriated for a tunnel from Brooklyn to Staten Island which was supposed to carry both freight and mass transit, which would develop another convenient dormitory for New York commuters. By this time almost all of Brooklyn was completely built up and Staten Island seemed to be the last frontier. Hylan's tunnel was begun in 1919. Shafts were sunk on either side of the Narrows. Then the project was abandoned for lack of adequate funds. The two excavations in Bay Ridge and Staten Island were dubbed Hylan's Holes.

At that point the New York and New Jersey Bridge and Tunnel Commission, which was in the process of building the Holland Tunnel between New York and Hoboken, began an investigation of the usefulness of building a tunnel or bridge between Howland Hook and Elizabeth and between Perth Amboy and Tottenville.

The commission issued a report in 1923 which stated that a span from Perth Amboy to Tottenville was "not only desirable and practicable, but that the lack of a fixed crossing between these points has in the past been the cause of great economic waste and has created intolerable traffic conditions which are steadily growing worse." This first official report called for immediate remedial action. A supplemental report by Clifford M. Holland, engineer of the Holland Tunnel, presented traffic estimates showing that by 1928, 12,162 persons and 2,696 vehicles per day would cross the Arthur Kill and the resulting tolls would pay for the bridge within 10 years.

Another recommendation was a combination railroad and highway bridge between Elizabeth and Howland Hook to replace the B. & O. structure. The cost was expected to be low, as the water was shallow and a low level bridge would solve the problem. However, the New Jersey State Board of Commerce attacked the idea of building a low level bridge and insisted that the bridge should be at least 135 feet above mean high water to insure the viability of New Jersey ports. The New York and New Jersey legislatures decided this was a reasonable demand, and they assigned the building of two high-level auto crossings to the Port Authority.

The Port Authority's involvement in the project caused a flurry among Staten Islanders. It gave rise to a movement by the locals to keep the work in the hands of the New York State Bridge and Tunnel Commission as the Commission had in the past concentrated its attention on the island's needs while the Port Authority had a great number of port projects among its activities. The movement against the Port Authority ended when the Authority issued a statement saying it regarded the building of the two bridges simultaneously as an integral port improvement. Port Authority traffic engineers explained that the two bridges would form an excellent bypass around traffic-ridden New Jersey industrial cities. Vehicles from northern New York State and New Jersey to southern New Jersey would undoubtedly save much time by using the two bridges. So these same activists turned their attention to getting the enabling bill passed by the New York State Legislature, for it had already passed in New Jersey. Enthusiasm in Staten Island for the bridges rose so high that on March 18, 1924, a special train carried to Albany a huge delegation from Staten Island to urge the passage of the bill.

Money for the bridges became available late in the spring of 1924. The Port Authority immediately increased its engineering staff and made such rearrangements in its organization as would carry forward the necessary studies, surveys, etc. Twenty-five percent of the funds were advanced by the states, so that the Authority could make more comprehensive surveys, and traffic studies to determine probable income from tolls from which prospective bondholders would be reimbursed. These studies, incidentally, turned out to be misleading. It was predicted that the two bridges would finish paying for themselves within 10 years of construction, while in reality it took the opening of the Verrazano-Narrows Bridge to bring enough traffic across the island to make the bridges self-sufficient. Although in the first year of operation the Goethals Bridge handled 674,500 vehicles and Outerbridge Crossing 511,000, the two bridges soon encountered the Depression of the 1930s and the gasoline rationing of World War II. It was not until 1945 that the Goethals Bridge handled as many as one million vehicles. Outerbridge hit the one million mark in 1947. Since the opening of the Verrazano-Narrows Bridge, each bridge carries several million vehicles a year. In the interim the Port Authority paid off bonds by pooling the income from other facilities.

When the Port Authority made its application for the two bridges to the War Department in the spring of 1925 it was discovered that certain shipping interests were strongly opposed to the bridges as planned, on the grounds that they would block water traffic and that the bridges would be "too wide" for boats to easily navigate beneath them. At the time, the New Jersey ports were doing booming business and the Arthur Kill and Kill Van Kull were handling as much tonnage annually as the Suez Canal. The Port Authority staunchly supported its designs for two cantilever bridges.

The Port Authority and various Staten Island and New Jersey civic groups began a campaign for the approval of the bridges by the War Department. Mass meetings were held on Staten Island, Elizabeth, and Perth Amboy to arouse public interest. Models of the proposed bridges were built to scale and exhibited in municipal buildings. The walls of meeting halls were covered with charts, diagrams and aerial photos of the Kills. Hundreds of people finally attended the hearings held by Colonel Herbert Leakyne of the War Department. After three days of debate Leakyne consented to the two cantilever designs. Soon the Port Authority issued and sold $14 million worth of bonds to underwriters.

COURTESY OF THE PORT AUTHORITY OF NEW YORK & NEW JERSEY

COURTESY OF THE PORT AUTHORITY OF NEW YORK & NEW JERSEY

Outerbridge Crossing, left, and the Goethals Bridge, above, are high level cantilever spans designed by J. A. L. Waddell. Outerbridge Crossing's main span is 78 feet longer than its rival's. Photo above also shows the Arthur Kill railroad lift bridge.

Outerbridge Crossing and Goethals Bridge are as alike as fraternal twins. They are both cantilever spans perched atop long graded viaducts. The land at both sites is very low, and in order to attain the required height of 135 feet above high water these viaducts had to be an impressive 10,800 feet and 8,600 feet long respectively. The weblike combination cantilever and truss span in the midst of the mighty Outerbridge viaduct seems almost an afterthought, a crowning decoration. Elizabeth Mock, in her aesthetic critique, *The Architecture of Bridges,* accuses the two cantilevers of "internal confusion—typical of an overhead truss with uneven upper edges." This is in reference to the maze of internal bracing above the members.

The Outerbridge Crossing consists of a 750-foot cantilever span, two 375-foot anchor arms, and a 300-foot through truss span at either end, making the total length 2,100 feet. There are 79 piers leading up to the span, each made of two concrete shafts connected by an arch. In all, about 34 million pounds of steel were used.

The Goethals Bridge is a shorter version of Outerbridge, with a suspended cantilever span of 672 feet and two 240-foot anchor arms. The approaches are not so long as Outerbridge, but are of the same character and have a maximum grade of four per cent.

Both bridges have four narrow lanes and a pedestrian walkway. Many drivers crossing the bridges complain of feeling dangerously constricted. It is not surprising, since the 42-foot wide roadways are divided into four lanes and two five-foot wide pedestrian paths.

Designer Waddell, an American, had experience designing structures throughout the world, but in a speech delivered to engineering students, he admitted that American engineers are untrained in architecture, and that to an extent this accounted for the unattractiveness of metal bridges in the U.S. Waddell also noted that American enterprise doesn't want to spend the money necessary to beautify bridges, and that competitive bids result in contracts to the cheapest design, regardless of talent and taste. Waddell said he considered himself guilty of geometric, perfunctory arrangements, but that he was attempting to refine the bridges he designed. Waddell is also famous for a primer he wrote for structural engineering students called *De Pontibus*.

In his efforts on Staten Island, Waddell had the aid of Professor William Burr of the Columbia University School of Engineering and of General George Goethals who had directed the construction of the Panama Canal. No difficulties were met sinking the piers, as the river bottom consists of silt to a depth of about 50 feet for the full width of the Arthur Kill. However, the Howland Hook approach required the use of piles.

Groundbreaking for the two bridges was a day-long affair. A coterie composed of Governor Al Smith of New York and Governor Silzer of New Jersey, Mayor Jimmy Walker of New York City and the mayors of Elizabeth, Keansburg, Perth Amboy and other New Jersey towns and the Borough President of Staten Island and scores more met in Manhattan and were ferried to Staten Island where they continued in buses and cars to the western side of the island where they made their first orations at Tottenville. Later they ferried across the Kills to the two New Jersey terminal sites. Returning to Staten Island, the group sat down to a formal dinner at St. George.

While New Jersey's ex-governor Silzer extolled the new links between Staten Island and the mainland that would transform the area, saying the "sleepy beautiful little places on Staten Island will come into their own," the *Times* scored Silzer's optimism, pointing to the devastation of Long Island's scenic charm as a result of improved transportation. The *Times* warned against the sacrifice of "beauty to land values which has ruthlessly eliminated trees, old picturesque houses and everything that could not be turned into immediate profits. Friends of Staten Island, while rejoicing in the prosperity that will come to it, cannot but hope that this will not be accompanied by the complete obliteration of its natural charm."

Records show that the bridges between Staten Island and New Jersey did not foster much suburban transplantation. Although the bridges were billed as a "backdoor" entrance to Staten Island from New York when used in conjunction with the Holland Tunnel, Staten Island remained the "Cinderella" borough. Although real estate interests kept advertising a boom in Staten Island, with New York and New Jersey residents arriving in droves, statistics show that the real net gain was a steady influx of about 4,000 persons a year, far less than the exodus to Queens in the same years. While the industrial district in New Jersey between the two bridges remains one of the densest in the world, for a long time grassy meadows and open fields lingered on the Staten Island side, evidence of the small farms and hamlets which once dotted the island.

The two Arthur Kill bridges opened to the public in June 1928, six months ahead of schedule. The opening ceremonies, held simultaneously on the two bridges, took place nine days before the facilities were opened to the public on June 29. The reason for this was that Governor Al Smith, one of the principal participants in the festivities, would be busy as a leading contender at the Democratic National Convention in Texas at the later date.

The southern bridge was named Outerbridge Crossing not, as many think, because it is so far from Manhattan. While it is possible that the Port Authority was engaging in puns, the crossing was named after Eugenius Outerbridge, the first chairman of the Port Authority and a Staten Island resident.

Outerbridge had joined the movement to keep the port intact when the towns of New Jersey were petitioning the Federal Government to be permitted to charge lower rates for freight delivered to the New Jersey side of New York harbor. He had been one of the drafters of the port compact which had created the Port Authority. He had urged the building of the bridges to Staten Island, for he believed that so long as Staten Island was isolated it was a hindrance rather than an aid to port facilities. One Port Authority official called the bridge "a monument to the foresight, sagacity and vision of Mr. Outerbridge." It was an unusual honor, for Mr. Outerbridge was alive when the bridge was named. Surprised by the public announcement, Mr. Outerbridge explained that his family name was derived from the fact that his ancestors lived near a bridge in England. He was a guest of honor at the opening ceremonies.

The Howland Hook Bridge had been designated the Arthur Kill Bridge, but General George Goethals, who had been the first chief engineer of the Port Authority, died shortly before the bridge was dedicated. It seemed natural to name the bridge after him instead. By an interesting coincidence June 29, the date of the opening of the bridge, marks the anniversary of General Goethals's birth. Goethals also was chief engineer on the Panama Canal and did some of the work on the Holland Tunnel.

The Port Authority decided to charge a toll of 40 cents per vehicle and an additional 5 cents per passenger crossing. But the ferry companies, which wanted to remain competitive with the bridges, put up placards on local street cars calling attention to the lower rates and shorter distance for motorists if they used the ferries. At the locations where the Port Authority had placed arrows showing directions to the bridges, the Staten Island Edison

COURTESY OF THE NEW YORK TRANSPORTATION ADMINISTRATION

Because the land on both sides of the Kills is so low, long high-level viaducts were required. Here work proceeds on the 8,600 foot viaduct for the Goethals Bridge. (1927).

Below, Staten Island Protestant Churches Convention wheels over Goethals Bridge two months after its opening in 1928.

Co. counteracted with arrows directing motorists to the ferries. So, as a final competitive move, the Port Authority lowered the bridge toll to 25 cents.

While these two bridges were being completed, the Port Authority decided to make studies for a bridge between Bayonne, New Jersey, and Port Richmond, Staten Island.

Although many schoolchildren are taught that the Sydney Harbor Bridge in Australia is the longest steel arch span in the world, in fact it is exceeded in length by New York's sweeping, elegant Bayonne Bridge, designed by Othmar Ammann. The Bayonne Bridge is all of 25 inches longer than the Australian bridge. The efficiency of the Port Authority, or perhaps its competitive spirit, is illustrated by the fact that the Bayonne Bridge was started five years after the Sydney Harbor Bridge, but was completed several months before its Australian counterpart. The bridge over the Kill van Kull may have been a conscious effort by the Port Authority to capture the steel arch bridge length record for New York, as such records are prized by cities as status symbols. However, many alternative plans were formulated before this arch was chosen. Studies of the three types of bridge suitable for a 1,500-foot crossing were made—cantilever, suspension and arch. They were all based on the same live loads, wind pressures, and temperature changes. The cantilever design proved to be the most expensive and least satisfactory in appearance, so it was quickly eliminated. Exhaustive studies were made of three types of suspension and two arch types. The cheapest type of suspension bridge was estimated at about ½ million dollars less than the arch, but the greater stiffness of the arch enabled it to carry the proposed rapid transit lines. This was the decisive factor.

Once the arch bridge was decided upon, more studies were made as to the type of arch suitable. The arch adopted has two hinges and increases in depth from crown to abutment. It consists of trussed ribs with a depth of 37½ feet at its crown and 67½ feet at the ends.

The Port Authority was still building Outerbridge Crossing, the Goethals Bridge and the George Washington Bridge when the studies began for the Bayonne Bridge. The inquiry into the bridge's financial feasibility had been completed by 1927. The report contained a series of recommendations for the bridge and gave an estimated cost of $16 million. The need for rapid transit facilities on the bridge was investigated by Ammann's staff. It was found that such facilities might be extensively patronized by commuters and others traveling between Manhattan and Staten Island by way of the Bayonne peninsula, but the revenue would be insufficient for some years to come to warrant the extra expense. It was decided, however, to build a bridge strong enough to permit the laying of rapid transit tracks in some future year, if necessary.

In fixing the exact location of the bridge, the determining factor was economy of approach. Topographically there was little need to give preference to one site over another. The final plans called for a skew bridge — built at a 58° angle to the water rather than one built at right angles to the water. This type of bridge is necessarily of greater length. Although this might be considered motivated by a desire to capture a span record, the reason given was that the skew bridge fit well into the scheme of highway connections on each side of the Kill van Kull. At this point the Kill was about 1,200 feet wide between pierhead lines with a channel of 30 foot depth (since deepened) running close to the Staten Island side. The bridge clears a height of 150 feet and its 1,652-foot span leaves an entirely clear waterway between pier lines. The Port Richmond approach is 370 feet, making a total length from plaza to plaza of about 2/3 of a mile.

Because the arch was built on a skew angle to the Kill it became a puzzle to non-engineers passing the site. The requirements laid down by the War Department said the channel of the Kill van Kull must be open to navigation at all times. That channel is located south of mid-span, so an unsymmetrical erection arrangement was compulsory. The south portion of the arch was erected extending for 14 panels and the north section extended for 26 panels. They were built up by hydraulic jacks and only a few very narrow steel rods were used as supports. The bents were removed as the arch sides neared one another. As the two sides got close—near the Staten Island side—the Port Authority was besieged by calls from commuters who passed the construction site daily. "You've made a mistake. The sides will never meet," the callers insisted. Nevertheless the two branches of the arch met within one inch. Closure was effected promptly and efficiently.

The design features a slender, slightly tapered hyperbolic curve over the roadway. The trusses of the arch are a pleasing repetitive pattern of regular triangles. The arch abutments were designed to have massive granite structures, but this plan was abandoned due to Depression cost-trimming. Many complain that this has left the Bayonne Bridge with an unfinished look, as the arch pins are exposed and its earthbound ends are not met with the solid connections they appear to require, but others feel this reinforces the purity of the arch structure. Despite the complaints from granite lovers, the Bayonne Bridge was awarded the prize of most beautiful steel ach bridge of 1931 by the American Institute for Steel Construction. The *Times* in a tribute to the new arch said there "is a symmetry and fineness of detail

COURTESY OF THE NEW YORK PUBLIC LIBRARY

Drawing of Bayonne Bridge from Richmond Terrace, S.I. is by Vernon Howe Bailey. There is a clear view of the arch abutments, which many consider "unfinished looking." Original plans called for granite facings on abutments.

COURTESY OF THE PORT AUTHORITY OF NEW YORK & NEW JERSEY

In order to preserve the navigability of the channel, only a few bents and a traveling crane were used to assemble the Bayonne Bridge. The 1,650-foot arch is set on a skew to the Kill Van Kull. Photo was taken in March, 1930.

about the Bayonne Bridge that is impressive and haunting."

The dedication of the Bayonne Bridge took place precisely three weeks after the dedication of the Gorge Washington Bridge. This fourth bridge brought to $143 million the expenditures by the Port Authority on its bridges and the purchase of the Holland Tunnel.

An international significance was given to the event by the participation of the Australian government, which sent its ambassador to present a rivet taken from the Sydney Harbor Bridge to the chairman of the Port Authority. The shears used to cut the roadway ribbon for the Bayonne Bridge were made of gold, and were sent to the Premier of the Australian state of New South Wales to be used at the dedication of the Sydney Harbor Bridge. Following that the shears were taken apart. One blade was returned to the Port Authority, the other kept by Australians as a memento of the two historic occasions.

D. M. Dow, secretary for Australia in the United States said that "J. J. Bradfield's design (the designer of the Australian arch) comprised an arch of a dimension making it the largest single span structure in the world. Then the engineers of the Port Authority came into evidence with a design for a longer arch, and so Australia was beaten by a few feet in the race for the distinction of bridge building. However, if the Bayonne span is longer, the Sydney structure is larger. Against the 16,000 tons of steel used in joining Bayonne to Staten Island, 37,000 tons were required to complete the structure spanning the beautiful harbor at Sydney."

A silver spike, a replica of the last spike used to tie the steel arch of Sydney Harbor Bridge, was given to the Port Authority chief. A telegram of felicitations was received from the French city of Bayonne.

COURTESY OF THE BROOKLYN PUBLIC LIBRARY BROOKLYN COLLECTION

Triborough lift span open to river traffic is seen through the lattice-work grid of the Willis Avenue Bridge. In distance is Hell Gate Arch.

TRIBOROUGH BRIDGES

MOSES' MIRACLES

The Triborough Bridge connects three land masses—The Bronx, Queens and Manhattan—by means of four overwater spans. It hops over two islands on the journey. The bridge, with its approaches, viaducts and overwater jumps, unwinds for 17½ miles. It was the first major bridge to link three areas.

The Triborough Bridge was not an experimental engineering success, as was the George Washington Bridge. It seems to have taken more political engineering than scientific engineering to get it built. Physically it is dominated by a 1,380-foot suspension bridge with trellis-like steel towers. Politically, it was dominated by Robert Moses, who maintained his leadership of the Triborough Bridge Authority despite a protracted personal feud with the Roosevelt Administration.

As the Triborough Bridge progressed from a dream spun by Harlem merchants to a steel and concrete reality, a parallel change was occurring in the city's political system. New York, during the years between the bridge's conception and its completion, changed from a starry-eyed expansion-minded city to one worried over preserving its autonomy and fighting to keep itself solvent.

The complications which arose in getting the bridge financed and the establishment of the Triborough Bridge Authority to solve this problem, set a precedent in city bridge building. The Triborough, or other authorities modeled after it, have built all the major bridges in New York since 1932. The subsequent authorities, all of which were headed by Moses, later were absorbed into the Triborough Bridge and Tunnel Authority, which in turn has been recently incorporated into the state-sponsored Metropolitan Transportation Authority.

Replying to a round of praise after the Marine Parkway Bridge in Brooklyn was completed in 1937, Moses, the sole member of the Marine Parkway Authority, admitted that he didn't yet understand the true nature of the five authorities he directed, "but they are great for getting things done." After all, Mr. Moses noted when the Triborough Bridge opened in 1933, "this bridge took 12 years of talk and two years of action." His history was a bit abridged. There were 21 years of talk and two years of action.

In 1904 Greater New York extended its subway north to Harlem. Population began to move with it and Harlem became an active community. Its merchants, who were congregated around 125th Street, advertised Harlem as the new center of New York. By 1913 several community groups began to petition the city for a bridge at 125th Street to Queens, saying it would build up the prestige of Harlem. Four impressive bridges now crossed the East River, but there was tremendous traffic overflow at 59th Street, for only the Queensboro bridge gave direct access to Queens. The congestion was so heavy that the Harlem merchants estimated a bridge built from Harlem to Queens would save New York merchants over $20 a million a year by redirecting and speeding up traffic.

By 1916 thirty-seven groups from the three boroughs had joined to petition the city for a new three-way bridge. Mayor John Hylan, a brash, blustering redhead who enjoyed little personal popularity, claimed that the city could not afford to build such an expensive bridge. A vocal foe of what he called the "traction interests," Hylan had initiated a city program to buy out privately owned subway and trolley lines, and to extend the lines to the new communities of Greater New York. Another of his pet projects was building a connecting tunnel from Manhattan to Staten Island. The proposed tunnel, combined with Hylan's determination to save the five cent fare on subways and trolleys through subsidies, would have cost the city about $130 million a year. The pay-as-you-go manner in which the city had financed the four East River bridges was impossible in this case and would drive the city bankrupt, Hylan maintained. The city's huge financial outlay for transportation already was causing it to exceed its $200 million debt limit. And the city was not allowed to issue extra bonds on non-revenue producing projects. The suggestion that tolls be collected on the new bridge met with skepticism. There was a legal hitch to it. After Mayor Gaynor had become a judge, he had ruled that the four East River bridges must be free, as they were really city highway extensions. So, it was generally accepted that charging tolls on a city bridge was against the law.

Despite its financial difficulties, the city gave the Department of Plant and Structure's bridge engineer, Edward Byrnes, a few thousand dollars to formulate some plans. Planning in conjunction with the local Harlem bridge advocates, Byrnes proposed a bridge which would cost $10 million, exclusive of approaches. The entrance was to be at the foot of East 125th Street.

After World War I ended and more money was in circulation, a bill to permit the building of the bridge was introduced into the State Legislature. The expense of construction was to be born by the city. Martin Healy, who introduced the bill, noted that the bridge would "rapidly become a great highway. It would develop a section of Queens that is only sparsely settled and add immeasurably to the value of real estate in Harlem east of Park Avenue that is declining in value, and would open a vast area in The Bronx."

Mayor Hylan, who had vetoed a direct appeal to the city for the bridge, was furious. He contended that the city did not need state permission if it wished to construct the bridge. Further opposition came from the Bronx Borough President, who said the bridge would serve only to give rich pleasure riders between The Bronx and Long Island a shorter route. It was more important, he felt, to build a bridge for direct relief of congestion on the Queensboro. The possibility of alternate sites was the basis of much opposition to the project. It was believed that a midtown connection to Queens, or another to Brooklyn at the Battery would have more effect in relieving traffic congestion on the East River bridges. Another suggestion was for a bridge from Ferry Point, The Bronx, to the Whitestone section of Queens. (These three alternatives all were realized later.)

Triborough Bridge advocates ignored these frustrations, however, and kept publicizing their cause, holding dinners and "snowballing" the issue. At one dinner in 1920, when the bridge was promoted as a "memorial bridge" for the dead of World War I, then-Congressman Fiorello La Guardia made short shrift of this ploy. An active proponent of the bridge, La Guardia declared he was tired of "having camouflage appended to public utilities. Let the bridge stand on its own merit as a means of communication between the three boroughs for the purpose of transporting passengers and freight, but please don't claim it to be a monument or a memorial. Every member who wants to put a pet project across thinks he could succeed by pinning it on the back of some dead soldier. Leave off the camouflage when you present the plan to the Board of Estimate. Just say that the proposed bridge is a needed public utility."

The original plans for the proposed structure called for a double deck span with concrete and steel pillars resting on Randalls Island. The upper level was to be for trolley cars and rapid transit, the lower for pedestrians, wagons and autos. The double arched granite towers resembled those of the Brooklyn Bridge. Nothing but arguments ensued. One constructive suggestion was offered by Gustav Lindenthal, the engineer who had just completed the Hell Gate Bridge in the same locale. Siding with advocates of the Harlem site, as opposed to an alternate East River crossing, Lindenthal suggested that spurs be built to the Bronx and Manhattan from his railroad bridge. A second deck would be built on the Hell Gate bridge for auto traffic. The suggestion, published in the *New York Times,* might have been an economical alternative to Byrnes' grandiose bridge. However, it seems to have gone unnoticed by city authorities, or perhaps it was vetoed by the N. Y. Connecting R.R., which owned the Hell Gate Bridge. Another suggestion that a private firm take over the building of the bridge got Mayor Hylan's temper going. Hylan warned the people of New York that if a private company built the bridge they could look forward to a "bridge trust owning nearly 50 city bridges instead of their being municipally owned." He likened this plan to the traction situation. To supress even the hint that a private firm should finance the proposed bridge, Hylan capitulated to the extent of appropriating a paltry $17,273 for preliminary planning. It was now 1925. The concession seemed auspicious enough to Harlem merchants to prompt a celebration at the Hotel Theresa followed by an enthusiastic parade. By this time the estimate for the bridge was upwards of $30 million, although engineer Edward Byrnes believed it might still be done for $15 million.

Natty, smoothtalking Jimmy Walker was sworn in as mayor of New York the following January. As a Manhattan State Senator, he had consistently supported the Triborough Bridge. He soon got the Board of Estimate to appropriate an additional $50,000. Optimism zoomed among the bridge advocates. However, Walker found city finances as dismal as Hylan had left them. As much as Walker favored the bridge, he could not come up with the entire sum from the city's coffers. He predicted it would be another 20 years before the bridge was completed if the city had to finance it.

One suggestion for financing the bridge was to raise the subway and streetcar fare from five to seven cents, and use the extra two cents for bridge construction. This idea was thoroughly blasted by the press. Commuters were outraged. And commuters were a large enough segment of the voting public to keep Walker from approving this measure.

But Walker and his Tammany cohorts were skillful at stretching the normal limits of the law. At the suggestion of Board of Aldermen President Joseph McKee, Walker petitioned Albany for permission to form a city bridge and tunnel authority modeled along the lines of the Port Authority, which could issue bonds to be repaid from tolls. This circumvented the strictures about raising the city debt limit, and would permit tolls to be charged on the bridge. However, Franklin Roosevelt, New York's governor, was puzzled about how to legalize this new intracity authority. It was necessary for the state to establish the authority, yet if the state ran it, would it constitute an invasion of home rule for New York City? And if the city ran it, was it just an obvious means to raise the city debt limit? A dispute arose in the state legislature.

While these legalities were being ironed out, Walker got the city to appropriate an additional $150,000 for preliminary soundings, hoping to speed up state deliberations. The following year, to give impetus to the movement to set up the authority, McKee convinced the city to appropriate $3 million to actually begin buying up land and building the anchorages. This act of good faith might encourage the state to take positive action, and could be paid back by the authority once it was set up, McKee noted. In addition, Tammany believed that the appearance of construction material on the bridge site would be highly beneficial to Mayor Walker, who was facing an election year. By now, 1929, more than ever, the population of Greater New York lived north of 125th Street.

Definitive bridge plans were published. The scheme immediately drew fire from professional engineers. Lindenthal, the most vocal critic, argued for a 116th Street entrance, which he said would cost less than the proposed 125th Street site, which was already too congested. Besides, the 125th Street location would spoil the setting of his own masterpiece at Hell Gate.

"Plans have appeared, I don't know by whom," Lindenthal wrote, "for a triborough suspension bridge of a cheap 'pole and washline' architecture located at Porter Avenue only 300 feet below Hell Gate Bridge. To build the bridge so close to the arch bridge would be an architectural outrage. One bridge would cover the other and the river landscape would be lastingly ruined. I trust that the aesthetic sense of the city authorities and of the public will not permit such a shame. It is possible for a new bridge to be located at 1,200 feet below the Hell Gate Bridge, a sufficient distance away to prevent one from blanketing another."

At the 116th Street location, Lindenthal said, the bridge would have a "distinguished architectural setting." He suggested that there, a well-proportioned steel arch bridge with concrete approaches would fit artistically much better into the land- and marine-scape. He also believed that from the major bridge two Manhattan connections should branch out: one four lane span across the Harlem at 102nd Street to give direct access to Central Park and a second at 116th Street. A third span at 125th Street could be added any time. Again he suggested his addition of a five lane second deck to the Hell Gate for Bronx-bound autos. If his suggestions were not heeded, Lindenthal warned, the Triborough would be obsolete within five years.

Lindenthal's objections were not egotistic rantings. His alternatives were fully backed by city master planner Robert Moses when he took office as Triborough Bridge Commissioner. Moses felt the connection at 125th Street was too far north and had been determined by the Hearst real estate interests without regard to the best interests of the city as a whole. (William Randolph Hearst owned some land at 125th Street on which the value was depreciating. He wanted the city to buy the land.) Moses noted that the bridge would be cheaper if the Manhattan crossing were further south, and in addition, it would prove to be a more convenient alternative route for Queensboro Bridge traffic if the bridge entrance were at 103rd Street. However, it was feared that the Hearst Empire would scuttle the entire bridge plan if Moses moved the Manhattan entrance, so, to make the Triborough more attractive to mid-town traffic, Moses convinced the Authority to extend the East River Drive northward from 96th Street.

Lindenthal's objection to placing the Triborough too close to the Hell Gate Bridge was also supported by the Fine Arts Federation, which petitioned the city to locate it as far from the Hell Gate as possible.

Further opposition to the bridge as planned came from the railroads and shipping interests objecting to the Harlem River crossing, which was to have a center pier swivel design. They claimed the 18-foot pier would obstruct and seriously menace car floats and tug boats, and that the obstruction in the water might cause increases in tidal velocities and be undesirable from the point of view of navigation.

Despite these disputes about the design, and despite still-unsettled problems in Albany concerning a new bridge authority, the Walker Administration blithely went ahead condemning and buying up land for the piers and handing out contracts for laying the cement foundation and anchorages in Queens and on Ward's Island. Governor Roosevelt, in a cooperative gesture, signed the Dunnigan-Reidy bill, which transferred state-leased land on Ward's Island to New York City so that the island could be used as a footstool for the new bridge.

Groundbreaking ceremonies were held Friday, October 25, 1929, the same day as the great stockmarket plunge, and a pointed two weeks before the next mayoral election. The city provided free bus and ferry service to Astoria Park, and throngs showed up from all five boroughs. Municipal bands supplied music and fire boats produced great sprays of water. There seemed perfect unanimity of opinion that Mayor Walker in particular and his administration in general were deserving of praise for their "courage, zeal and foresight" in getting the bridge project underway.

"Beau James" Walker, who secretly hated crowds, although crowds loved him, showed up with silver spade and silver tongue to turn the first shovelful of earth. He gingerly scooped up some dirt from a grassy slope while photographers jumped around him. Walker commented with a grin that this was his "first attempt at mudslinging," and continued working with his shovel. When the noon whistle blew and an official suggested it was time to stop for lunch, Walker retorted, "We shall not knock off for lunch on this job." A few moments later, however, tired by the physical exertion and exasperated that the crowd was following him around as if it were a golf match, Walker laid down his tools, quipping that he had "not agreed to continue manual labor until the entire job is completed." Two weeks later Walker won a sweeping victory. The Triborough, however, crept along until the $3 million was spent, its only material achievement an anchorage at Ward's Island, completed in 1932. The project was then abandoned, a victim of the Depression.

The *Times* pointed out the sharp contrast between the huge lump of cement on Ward's Island, the only visible evidence of the Triborough Bridge, and the rapid erection of the George Washington Bridge on the other side of Manhattan Island. To quiet this criticism, a Triborough engineer publicly replied that the Triborough was a much more complicated structure from an engineering point of view than the George Washington, for the George Washington was a single span while the Triborough was to include three separate bridges, a long viaduct and a complicated traffic intersection to eliminate all cross traffic. However, it was rather obvious that work had been suspended. Employees were laid off. The Walker Administration was in trouble. The *Times* denounced the tactic. "While the city is spending millions of dollars to provide for its army of jobless, it is augmenting the number of unemployed by refusing to continue work on the bridge. The estimated tolls would bring in an income to the city of $2¼ million a year." But the city was unable to come up with more money for bridge construction. The cost of running Greater New York had risen from $92 million in 1900 to $631 million in 1932, and even Mayor Walker couldn't pull more money out of his high silk hat.

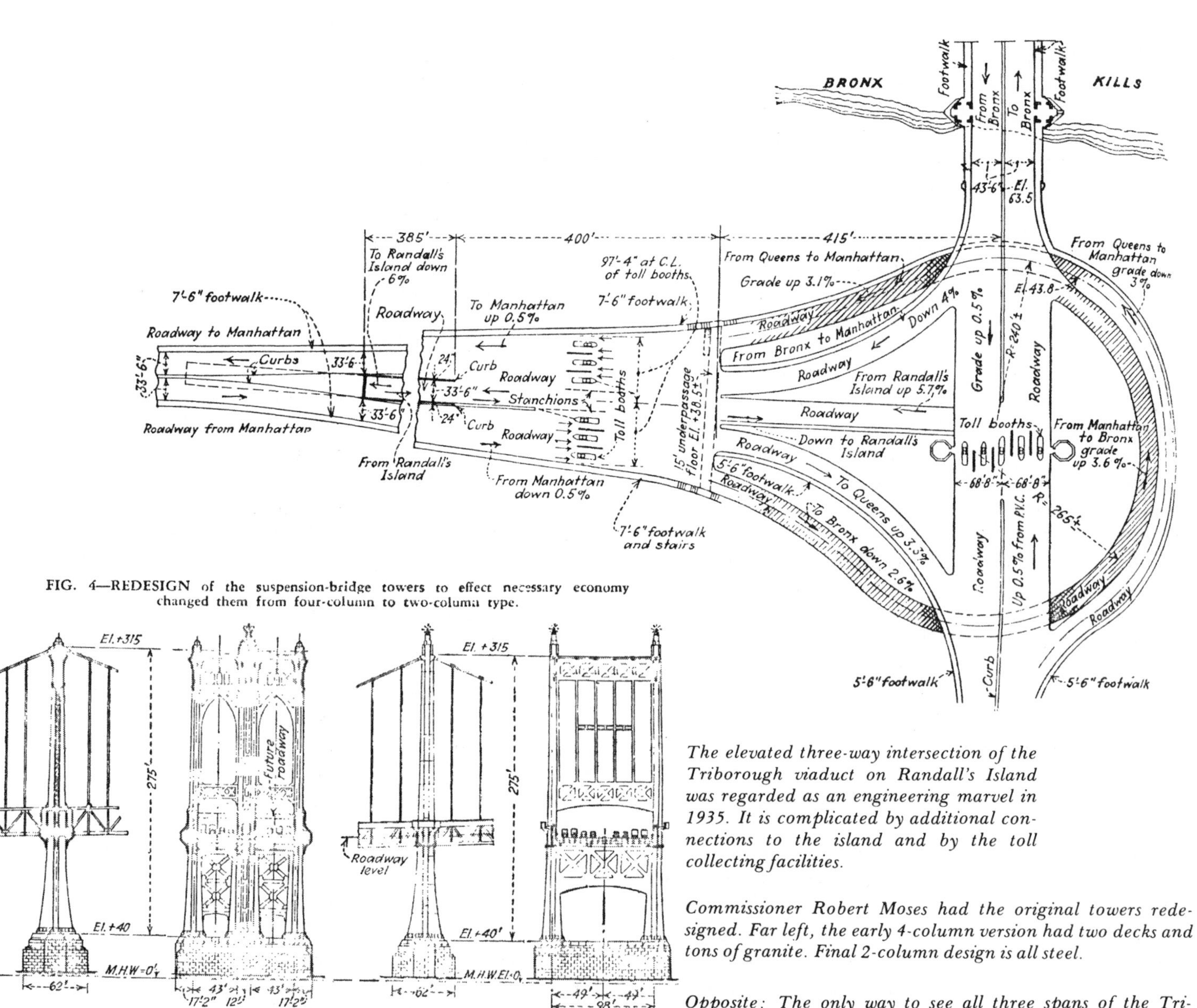

FIG. 4—REDESIGN of the suspension-bridge towers to effect necessary economy changed them from four-column to two-column type.

The elevated three-way intersection of the Triborough viaduct on Randall's Island was regarded as an engineering marvel in 1935. It is complicated by additional connections to the island and by the toll collecting facilities.

Commissioner Robert Moses had the original towers redesigned. Far left, the early 4-column version had two decks and tons of granite. Final 2-column design is all steel.

Opposite: The only way to see all three spans of the Triborough Bridge at once is from a plane or helicopter.

Later that summer the Hoover Administration set up the Reconstruction Finance Corporation. Robert Moses, Chairman of the State Council of Parks, immediately applied for Federal grants. Various civic groups urged the city to borrow money to construct the Triborough. They reasoned that the 38th Street tunnel and the Triborough Bridge were probably the most necessary improvements of the type covered by the provisions of the RFC in the United States. In fact, the tunnel was mentioned during the debate in the House and Senate as an outstanding example of the sort of public undertaking which could be financed by such loans.

Senator Robert Wagner, at Walker's suggestion, applied to the RFC for a $37 million loan. The application was "in the mill" when Walker quit in the midst of an investigation of his administration. The new acting mayor was Joseph McKee, a staunch fiscal conservative. McKee opposed the application to the RFC because he considered it a confession of municipal bankruptcy. Anyway, he noted optimistically, the city's bonds would be in fine condition by December and then the city could borrow all it needs locally without recourse to the Federal Government. McKee claimed this was only one of those cyclical depressions and was bound to change soon. "New York is able to care for its own needs without a cry for help. The evils that would flow from such an appeal would far outnumber the benefits of putting a few men to work. The question arises, where would such a program stop? If we borrow money for the Triborough Bridge why not for the subways? In this crisis we are getting far away from American principles." Despite arguments from men in his own party, including former Governor Al Smith, who called McKee's principles "high hat," McKee managed to block the RFC loan. The following year a new mayor, John O'Brien, was elected.

COURTESY OF THE TRIBOROUGH BRIDGE AND TUNNEL AUTHORITY

O'Brien investigated city finances thoroughly and found them in shambles. He suggested an old solution for building the bridge: the creation of a municipal bridge authority modeled after the Port Authority to complete the Triborough through issuance of bridge authority bonds. He gained the support of Al Smith. This Authority was not merely conceived as an answer to the problem of building the Triborough Bridge, but later might assist in building the Henry Hudson Memorial Bridge and a connection to Staten Island, O'Brien maintained.

The new Roosevelt Administration agreed to approve a loan either to the new authority or to the city. By April the Triborough Bridge Authority was approved in Albany as a method of putting thousands back to work. The vote was unanimous in favor of the bill, and the architect of the bill was a maverick Republican who simultaneously held three jobs under New York State's Democratic Administration, and was known as the best bill drafter in Albany: Robert Moses.

The name Robert Moses is inextricably linked to the achievements of the Triborough Bridge Authority. Unlike his biblical namesake, Robert Moses got his travelers across New York's waterways by bridging them, not parting them. However, there are those who say the present-day Moses believed himself as righteous as the Old Testament leader; others admire his fortitude, his public service, his commitment to technological progress combined with humanistic amenities. After all, the bridges of the Triborough Bridge Authority were surrounded by swimming pool complexes, parks and picnic areas. Moses continues to be one of the most controversial personalities in contemporary New York politics. His arrogant confidence and highhanded tactics have made him many enemies since the beginning of his career, but his effectiveness has earned him more titles and public jobs than any other man in American politics.

Despite Moses' involvement in drafting the Triborough Bridge Authority bill, he was not chosen as one of its commissioners. Mayor O'Brien selected three Tammany men — Fred Lemmerman, Nathan Burkan and John O'Leary.

Roosevelt's Public Works Administration, directed by Secretary of the Interior Harold Ickes, granted the Authority a \$37 million loan, to be repaid in 20 years with an interest rate of 4%. It was advanced against a first lien on revenue derived from the bridge. Mayor O'Brien proudly announced that "there would be no delay in resuming operations on a large scale." The Triborough Bridge was the first non-Federal city project to earn PWA approval.

The three commissioners were immediately besieged by job hunters, but the PWA stipulated that these men be civil servants. By the end of the year the hiring had begun, but work was still delayed by red tape.

The next year still another mayor took over. This time it was the 'Little Flower," Fiorello LaGuardia, a short, outspoken man "of the people," who seemed almost naively uncorrupt. However, LaGuardia could be ruthless. LaGuardia wanted to gain control of the Triborough Bridge Authority because he doubted the honesty of O'Brien's Tammany appointees amidst so much money. Sixteen days after taking office, LaGuardia exposed an arrangement by Commissioner Lemmerman through which the Triborough Bridge Authority paid office rent to his firm. He also accused Joseph Johnson, the Authority's general manager, of impeding and delaying the work. "These bridges are to be built of steel, not s-t-e-a-l," LaGuardia told newsmen. Lemmerman resigned, but as a result the Federal Government decided to withhold funds for the bridge until the ruckus was settled.

The authority's chairman, Nathan Burkan, claimed that LaGuardia's accusations were unwarranted and despicable. However, LaGuardia found evidence that Commissioner O'Leary had been referring engineers seeking jobs with the authority to Bronx Tammany leader Edward Flynn. By February, Joseph Johnson retreated under fire and resigned his \$13,500 job — a considerably higher salary than city engineers received. O'Leary resigned soon afterward.

LaGuardia appointed George McLaughlin, a banker and reform Democrat, to replace Lemmerman. Moses got O'Leary's job. Now the Authority represented the city's fusion administration, with a Tammany Democrat, a Liberal Democrat and a Republican.

COURTESY OF THE BRONX COUNTY HISTORICAL SOCIETY

The main suspension span is 1,380 feet long and its towers rise 315 feet. There are 10,000 tons of steel in the suspended structure.

The Moses appointment received special acclaim. Moses was already City Parks Commissioner, Chairman of the State Council of Parks and President of the Long Island State Parks Commission. A special act of the State Legislature enabled him to hold his state posts at the same time he served the city. The mayor assumed that the legislation would cover the extra position.

Byrnes was soon dismissed from his $20,000 a year position as chief engineer of the Triborough Bridge Authority. The original plans for the Triborough were in the deluxe manner of the Walker Administration. Byrnes had approved a design including four cables, two decks with 16 roadways and a granite facing. The four stiffening trusses were to be 24 feet deep and the elegantly decorated towers were to consist of four columns. In keeping with Moses' philosophy of economy, efficiency and comprehensive planning, Moses called in Byrnes within a fortnight of his appointment and asked the engineer what he thought more important: adequate approaches or ornamental granite. When Byrnes unhesitatingly replied "granite," he was told to resign and get his pension. Moses then called in the Port Authority's staff of engineers to draw up new plans. Othmar Ammann and his men were already in close touch with the latest developments in bridge building, having just finished the George Washington Bridge. Under Ammann's direction, the Triborough design was immediately cut from two decks to one and from 16 lanes to eight. Engineers' estimates, Moses said, "have put the time when 16 lanes will be necessary to 30 to 40 years from now. By that time we should be in a position to afford a new bridge." All the granite facing proposed (almost 2 million cubic feet) was eliminated. The granite had been designated as a boon to friends of Tammany. Of the original 17 concrete piers already built on Ward's Island, the new design utilized 13. Also the panel lengths of the stiffening trusses (now 20 feet high) were increased so fewer suspender ropes were needed. Altogether the estimated cost was reduced by $2 million.

Although LaGuardia was now satisfied that the bridge would get underway swiftly with Moses leading the operation, the Roosevelt Administration was disgruntled. Moses had tangled with Roosevelt when the latter was Chairman of the State Council of Parks and Moses was one of the commissioners. The conflict originated when Moses and Roosevelt split on how to spend the slender funds of the newly consolidated commission. Roosevelt preferred a parkway to the partially developed Taconic Parks area in upstate New York, while Moses favored one to the Long Island parks. Both were stubborn men. Moses convinced Governor Al Smith that the Long Island park and parkway system should have priority because New York City dwellers were in greater need of recreational facilities than their upstate neighbors. Roosevelt resented this.

But the real schism between the two men began when Roosevelt went about seeking political office.

He proposed an Albany journalist named Louis Howe be appointed secretary of the Taconic Parks Commission. Moses, learning that Howe was really pegged to serve as Roosevelt's campaign secretary and didn't intend to devote any time to the commission, became infuriated. He considered the appointment a matter of political ethics and determined to fight it. As Moses' notoriously tactless temper flared, he told Roosevelt that if he wanted a "secretary and valet" he would have to pay for them himself, as no sinecures could be allowed in the state park commission. The future president never forgave him. The conflict between "battling Bob Moses" and the aloof but vindictive president was a paradigm of political gamesmanship. For the first year of Moses' stint with the Triborough Bridge Authority, Roosevelt tried to oust Moses through personal influence and legal maneuvers. Moses stood his ground.

The first public inkling of the Roosevelt Administration's displeasure with Moses occurred when Federal checks for bridge costs failed to arrive in March, 1934. Privately, city officials had been warned the preceding month that the Administration desired Moses' resignation on the ground that he was not sympathetic to New Deal philosophy. Moses, meanwhile, continued to send out glowing descriptions of the new bridge. The $2 million to be saved by eliminating the superfluous decorative touches was to be used for improving bridge approaches. "Beside the Nile the anchorages would pass for pyramids" and the 390-foot-tall towers of the suspension section were "tapering, graceful, ornamental." Funds stopped flowing altogether by July. By the end of the year contractors were pressing the Public Works Authority for payment and Moses admitted to the press that it was "a serious situation, and the effects on the credit of the Authority are just as bad as if the debts were large, for these contractors will tell others and the result will be that no one will bid on our works, or if the bids are submitted they will be very high."

Mayor LaGuardia received a telegram from the Roosevelt Administration after the 1934 gubernatorial election, which Moses had lost to Democrat Herbert Lehman. It warned the mayor that if Moses were reappointed when his term as Triborough Commissioner expired, all public works funds would be withdrawn from New York City. LaGuardia, although sympathetic with the New Deal, refused to be bullied.

By January 1935 the dispute reached the front page of the *New York Times*. Under headlines reading "Moses Bridge Job in Peril as Ickes Seeks his Ouster: LaGuardia in dispute with PWA head as latter withholds Triborough Bridge Funds," the *Times* disclosed a new Administration tactic: the damning text of Administrative Order 129:

"Hereafter no funds shall be advanced to any

COURTESY OF THE BROOKLYN PUBLIC LIBRARY BROOKLYN COLLECTION

Workers begin spinning the 10,400 miles of wire that form the cables on the Triborough suspension span. There are two cables 20-5/8 inches in diameter. They are placed 98 feet apart.

authority, board or commission constituting an independent corporation or entity created for a specific project wholly within the confines of a municipality, any of the members of the governing body of which authority, board or commision holds any public office under the said municipality."

Despite its impersonal tone, this Administrative Order, preventing one man from concurrently holding a municipal and PWA-project office, applied to only one man and one authority in the country: Robert Moses and the Triborough Bridge Authority. When Moses and LaGuardia leaked the order to the newspapers, Moses claimed it was a "vendetta" and that "if the outlay of public funds were to be governed by personal and political reasons, the public should be informed." Ickes immediately denied that the administrative order was aimed specifically at New York City. However, many years later, in an article called "My 12 Years with Roosevelt" for the *Saturday Evening Post,* Ickes recanted, telling the story of Roosevelt's grudge against Moses, and how it was dumped in his lap. Ickes insisted that this was the only instance in which he had ever attempted to accomplish a "political result for F.D.R." and that he had never heard of Moses before Roosevelt's orders came through.

Moses also publicized his reaction to the Administration's demands. He said that when the mayor informed him of Ickes' attitude, he offered to resign both his Federal and local positions if the mayor so desired, but he refused to continue as Parks Commissioner while taking a "backdoor exit" from the Bridge Authority.

Roosevelt's action aroused protest from civic groups throughout the metropolitan area. Everyone who knew anything about the Triborough Bridge project knew that Moses was the only one of the three commissioners who devoted much time to it. The protests began pouring in, but on January 9th Ickes issued what amounted to an ultimatum, declaring he was waiting patiently for Robert Moses, New York Parks Commissioner, to give up his post on either the Triborough Bridge or the City Parks Commission. Ickes said that the aim of the Administrative Order was to properly separate the Federal and local governments. He denied that President Roosevelt was responsible for the uproar.

By January 16th, 147 organizations in Greater New York published a long letter accompanied by a brief calling Order 129 illegal and void, "an arbitrary and capricious fiat without any authority in law." After this first public expression of disgust, a vast number of letters and petitions flowed into Washington, asking that Moses be permitted to retain his position in the bridge authority. The farflung praise Moses reaped in this period is exemplified by a letter from the Auto Club to Ickes saying:

"His unquestionable honesty and altruism, his boundless energy is always unstintingly and courageously directed for public welfare, and his application of efficacious yet idealistic methods in every job he tackles are indubitably facts known and readily admitted by every right thinking New Yorker regardless of party, creed or religious inclination."

By March, Washington was inundated by letters and telegrams from pressure groups and individual citizens. Except for one union which accused Moses of union busting tactics and threw its support to the Administration, all groups expressed approval of Mr. Moses' performance. LaGuardia spent January and February practically commuting to Washington to iron out the dispute. Fearful that the Public Works Authority would cut off all relief funds from the city, LaGuardia consistently issued statements that there were no arguments between him and Washington.

Finally, Roosevelt decided to capitulate. As he had never publicly stated that Executive Order 129 was specifically directed against Moses, there was a simple face saving device. Two predated letters were issued. One, by Mayor LaGuardia, pointed out that Order 129 affected the reappointment of Langdon Post as Chairman of the New York City Housing Authority, with a postscript saying the Order also involved Mr. Moses. In another predated letter, Ickes wrote LaGuardia that he would interpret Order No. 129 as not being retroactive. Thus, Moses and Post retained their several positions, and Mayor LaGuardia humbly promised never to make such duplicate appointments again. The *Times* referred to the tactic as "complete surrender."

Throughout the standoff, Moses kept confidently issuing statements that the bridge would indeed be finished on schedule—July 1936. Alternately he remarked that the Federal Government's withholding of funds would be the only reason the bridge might not be completed on time. Moses went on spending money, both appropriated and not appropriated, banking on his ultimate victory.

But once the battle was won, Moses began another campaign. This time he set out to create a network of bridge approaches and parkway connections which would make the Triborough Bridge an integral part of the city's recreational facilities. As New York City Parks Commissioner, Moses was in a position to integrate all these facilities. "The day of building crossings from plaza to plaza and dumping traffic on the municipalities is over," Moses proclaimed. He believed the Triborough Bridge should be a part of a grand scheme to make the city more livable. As the Triborough was a complicated system of crossings, its approaches would provide a systematic linking of the city's parks. Thus, Moses convinced city and state to create 14 miles of new highway approaches for

COURTESY OF THE BROOKLYN PUBLIC LIBRARY BROOKLYN COLLECTION

This was the longest highway lift bridge in the world when the Triborough was built. Altogether it is 770 feet long and its lift portion is 310 feet and weighs 2050 tons. Here it opens for the first time. The crowd gathered for the occasion included Mayor Fiorello La Guardia and Commissioner Robert Moses.

COURTESY OF THE BROOKLYN PUBLIC LIBRARY BROOKLYN COLLECTION

Crane hoists into place members of the 20-foot high stiffening truss. Plate girders on roadbed are 96 feet long.

the bridge, stretching from Pelham Bay Park in The Bronx to the Flushing Meadows in Queens. Included in the scheme were parks on Ward's and Randalls Islands, and a stadium on Randalls Island; the extension of the East River Drive from 96th Street northward to entice downtown traffic to the new bridge; a yacht basin on Flushing Bay; a park with an Olympic size swimming pool in Astoria.

While these improvements were springing up, a fascinating variety of spans was materializing between the three boroughs. Spreading out like a Y from the Randalls Island toll junction—one of the most complicated traffic junctions ever designed—the bridge included a series of viaducts, a suspension bridge, the longest lift bridge in the world, depressed roadways, ramps, and, of course, the elaborate junction. There is no single location along the bridge where all its members can be viewed as a unit. One can only catch sight of a link here and there—a whole bridge, parts of two others, stretches of a seemingly endless viaduct.

The lift span, across the Harlem River, has a boxy, structural look. Its lift span is 310 feet long and weighs 2,050 tons. Within each of its twin 215 foot tall steel towers is a thousand ton cement counterweight and a 200 horsepower motor capable of hoisting the six lane lift span and its supports 80 feet higher than the span's regular closed position or 135 feet above the Harlem River.

Over the Bronx Kills is an eight-lane truss bridge with a main span of 383 feet connecting Randalls Island with The Bronx. This bridge has been designed so that it may be converted into a lift bridge if it ever becomes necessary.

To the observer, the most prominent feature of the Triborough Bridge is the 1,380 foot long suspension bridge between Randalls Island and Astoria, Queens. It is a two cable affair which carries eight auto lanes. The 98 foot lateral spacing between cables on the bridge was exceeded only by the width of the pairs of cables on the George Washington Bridge. The flexible steel towers are designed for a cathedral-like effect. Along all three spans there are pedestrian walks, connecting The Bronx, Queens and Manhattan and providing access to the Randalls Island park area. Another remarkable feature is the length of the steel floor beams. The great width of the 8-lane roadway required plate girders 96 feet long and 8 feet 4½ inches deep.

The heart of the monster, and the most significant engineering feature of the bridge, is the section where the routes join in a serpentine concrete arrangement on Randalls Island. This junction contains the toll booths. Considered the most ambitious traffic mechanism ever built, it was modeled after the "flying junctions" used by railroads. Its three decks are designed to allow its 22 roadways to converge and to radiate but never to cross. Motorists from any of the three boroughs get to their destination after passing only one toll booth. According to Ammann, the project combines "speed and safety as well as a form of recreation to the traveling public to an extent not equaled in any similar project."

But the engineering considerations of the new bridge were overshadowed by the personality and comic political operas attached to Moses' commissionership. In 1935 Moses, scheduled to be reappointed as commissioner in July, began hearing rumors that LaGuardia had made a secret deal with Roosevelt which would relieve him of his Triborough job when his term expired. As July approached, the mayor compounded Moses' suspicions by stating he believed it unnecessary to reappoint Moses formally,

COURTESY OF THE MUSEUM OF THE CITY OF NEW YORK
The 125th Street approach of the Triborough Bridge in June 1937.

but rather that Moses would retain his post as "holdover." Moses, concerned that this holdover status might be a political trick, insisted that unless a formal swearing-in ceremony was staged he would leave. LaGuardia called Moses a prima donna and a baby, but finally gave in.

Moses' last quarrel with the Roosevelt Administration came in the last months of work on the Triborough Bridge. In April 1936 Moses requested that laborers be permitted to work a 40-hour week on the Randalls Island junction. His request was denied. The normal PWA work week was 31 hours, and Moses' project was to be no exception. Moses kept requisitioning the PWA. He claimed his request would combat a labor drain—because all the competent workers were being lured away by private competitors who offered a longer work week.

Moses again eventually got his way, and the 40-hour work week was declared permissible in June 1936—one month before the scheduled completion of the bridge. This go-ahead was accompanied by a voucher from the PWA authorizing $15,000 in Federal funds to be used for entertainment for the opening celebration. An accompanying statement announced that President Roosevelt himself would make an address at the opening. Ickes declared that he would not attend and that "outside of the special invitation to finance the party, I received only one of those general invitations broadcast to everyone in the city."

Saturday, July 11, 1936, the opening day of the Triborough Bridge, was the fourth day of one of the most blistering heat waves the nation had ever witnessed. Yet workmen were up all night and into the next morning paving unfinished sections of the roadway and removing debris. Over 700 deaths had been reported nationwide as a direct result of the heat. The New York area alone accounted for 72 of these fatalities. Nevertheless, 15,000 souls braved the sun, walked over the new spans and stood below a ramp leading to Randalls Island and the new Municipal Stadium, where the Olympic trials were taking place. On a platform, in a stylish white suit and Panama hat, stood the President, accompanied by Governor Lehman, Mayor LaGuardia, Robert Moses, and Harold Ickes, who had received a last-minute invitation from the Mayor. As a special request, Robert Moses was not to introduce the President, though they did occupy the same platform and Moses introduced the rest of the speakers.

As the President's motorcade proceeded to this formal bridge dedication during an election year, thousands lined the streets of Harlem to catch a glimpse of the man who was hailed and cheered as the savior from the Depression. At the ceremonies, Roosevelt defended his policy of financing public works and of using Federal revenue on local projects. "People require and people are demanding up-to-date government, just as they are requiring and demanding triborough bridges in the place of ancient ferries," Roosevelt said in a speech that was considered generally unpolitical. Mayor LaGuardia reiterated Roosevelt's more diffident defense of spreading public funds, declaring "we dedicate ourselves to the building of a greater bridge which will permanently join the land of liberty and equality to a system of economic security."

Harold Ickes proclaimed the bridge as one of the most impressive works undertaken by the Federal government under the auspices of the PWA, and catalogued other projects the Roosevelt Administration had financed in the metropolitan area. Neither the President nor Ickes mentioned Moses' role in the colossal project.

Moses' and Ickes' sharing of the speakers platform brought amused smiles to the faces of politicians and journalists. The two had never met before. Ickes and Roosevelt ignored Moses in their speeches, although Governor Lehman and Mayor LaGuardia praised him. However, Moses was not one to let things go unspoken. Referring to the President and Ickes taking credit for the bridge, Moses quoted from a letter Samuel Johnson sent to Lord Chesterfield, in which the great English scholar, after finishing his dictionary, said to the lord who claimed to be his patron that he had finished it "without one act of assistance, one word of encouragement, or one smile of favor. Such treatment I did not expect for I never had a patron before." Moses praised this letter as one of the "finest pieces of polite vituperation in the annals of English Literature," and implied that it had obvious parallels to his own relationship with the PWA. "However," Moses concluded, "I have always reflected that after all, the important thing was the completion of the dictionary, not the writing of the letter."

Despite his "no grudge" attitude at the ceremonies, Moses had no intention of getting involved in Federal intrigues in his future projects. He had already created the Henry Hudson Parkway Authority and the Marine Parkway Authority, which were busy building bridges. He was now contemplating the Bronx-Whitestone span, which was to be undertaken by the Triborough Bridge Authority to provide a route for tourists to the proposed World's Fair of 1939. The fair was to be held in the newly resuscitated meadows at Flushing. The go ahead on the financing of the new bridge depended only upon the financial success of the Triborough.

After opening ceremonies, 200,000 people rushed to cross the Triborough by auto, bus, cycle and on foot. The Presidential party was the first to drive over the span, but later all approaches were jammed with enthusiasts waiting to cross the new bridge. The largest police contingent since the Lindbergh parade stood ready at the bridge entrances. The word to open the bridge to toll traffic was flashed from a special shortwave field station to police radio cars and motorcycles. On receiving the signal, the barriers at the Queens and Bronx entrances were lowered and the huge lift span to Manhattan began slowly to descend. Private automobiles, taxis, motorcycles and even a tandem bicycle raced forward, each eager to gain the honor of being the first paying vehicle across the bridge. Although it looked, in the heat of the race, as if a fellow on bicycle would be the first to pay the toll, it turned out that in the excitement he had forgotten to bring a dime. Next to arrive at the toll gate was Omero Catan, a man who made a hobby of being "first" (he had been the first on the George Washington Bridge). Catan had been waiting since 2 a.m. at the Manhattan entrance. In the early morning, however, after measuring a map, he had decided the Bronx entrance was closer to the toll booths and moved there. When he was 100 feet from the toll booths Catan's car stalled. He raced forward on foot to give the toll to the employee in the booth. The attendant refused to take Catan's money, since he had no vehicle. Sympathetic press members and onlookers helped Catan push his car forward.

The estimated capacity of the Triborough Bridge was 60,000 automobiles a day. Tolls collected the first year were sufficient to repay the Federal loan, a good indication that the proposed Bronx-Whitestone Bridge would also be a money making proposition. Thus, private bankers were soon induced to buy Triborough bonds for the new suspension bridge.

By 1934, Moses had already convinced private banking groups to buy special authority bridge bonds from the Henry Hudson Parkway Authority and the Marine Parkway Authority. He was the sole commissioner in both of these projects. "I did not collect hats and badges, as some unkind critics have charged, for the sake of wearing them, but to insure coordination for a number of complex interrelated undertakings and to eliminate or at least reduce the petty jealousies, rivalries and complaints so common in government work," Mr. Moses maintained in a book he wrote called *Public Works: A Dangerous Trade.* By 1936 Moses was supervising construction of three bridges as well as a number of highways and parks.

The Henry Hudson Bridge over Spuyten Duyvil was built as part of the Henry Hudson Parkway Improvement. A bridge had been previously proposed to commemorate the Hudson-Fulton celebration of 1909 and a design for a concrete arch was submitted. However, the only permanent structure the Hudson-Fulton Committee did erect was the shaft of a monument to Henry Hudson on which a statue of the navigator was to be placed. Then the entire project was abandoned as the Art Commission turned down the design, judging that the classical arches were unfit for the forest cliffs. Although succeeding city administrations backed a bridge at the site and signs

Scientific American *featured this design for a Hendrik Hudson Memorial Bridge on its cover in June 1906. It was to be part of the Hudson River tercentennial celebration of 1909. The city's Art Commission jettisoned the plan.*

Henry Hudson's arch keystone is about to be lifted into place from anchored barge below. Photo was taken from The Bronx.

were even posted to announce the imminent building of a bridge, no funds were ever appropriated.

None of these plans suggested that the bridge be linked with an entire West Side improvement. Like other Harlem River bridges, it was assumed that this bridge, too, would lead to a city street. It took Robert Moses' vision to see the bridge as part of a huge West Side improvement plan that would realize another long vaunted but never begun project for Manhattanites. This improvement included relocating and covering the New York Central tracks, which ran along 11th Avenue. (This street was called Death Avenue, and featured cowboys on horseback waving red flags as they rode ahead of the slow moving trains.) A plan to eliminate the tracks had been agreed upon by Manhattan Borough President Julius Miller and the New York Central in 1929. At that time the construction of the West Side Express Highway began. However, above 72nd Street, the uncovered tracks, with what Moses called "miles of smelling cattle trains" were a prominent feature of Riverside Drive and the ribbon park which adjoined it.

Moses put together a variety of projects that included enlarging the park by covering the railroad tracks, building a yacht basin at 79th Street and building the Henry Hudson Parkway and Bridge. As parks commissioner, Moses secured funds from the Parks Department. He also received grade crossing elimination money from the New York Central, and to finance the bridge he sold $3.1 million in bonds to a group of private financiers. The bridge

When joined, Steinman's arch became "hingeless."

would be an arch type and would be designed by David Steinman.

Actually, Steinman's steel arch bridge for Spuyten Duyvil had been on the drawing boards for years. Steinman was a graduate engineering student at Columbia University in 1909 when the Hudson Memorial Bridge had first been suggested. Although an established engineer got the project, Steinman designed a fixed steel arch as a student project for his civil engineering degree. It was very close to the design Steinman submitted to the Henry Hudson Bridge Authority in 1934. The fixed or hingeless arch takes advantage of the high rocky cliffs along the Harlem by transferring all the thrust from the arch to the cliff walls. Steinman's plan called for a six lane arch bridge of 800 feet with two 300-foot side spans, but the financiers were unsure that they wanted to back such a large bridge. Although the highway leading to the bridge had six lanes and traffic studies seemed to insure them of a good return on their investment, they felt that since all the other Harlem River bridges were free, the 10 cent toll on the Henry Hudson Bridge might make drivers avoid the new bridge. The bankers insisted that Steinman

Steinman had to add a second deck to the bridge in 1937. Although this photo shows the bridge in aluminum paint, it was originally dark green to blend with Inwood Hill Park.

COURTESY OF THE TRIBOROUGH BRIDGE AND TUNNEL AUTHORITY

Below, "Century" rig sets beam on reinforced concrete piers of Cross Bay Bridge in 1958. This modern bascule replaced a low level viaduct and bascule built in 1938. (For photo of earlier Beach Channel bridge, now demolished, see page 139.)

COURTESY OF THE TRIBOROUGH BRIDGE AND TUNNEL AUTHORITY

slim the bridge down to four lanes. Steinman agreed, but suggested the bridge be made strong enough to carry an extra deck.

Steinman's life was a love affair with bridges. Born on the Lower East Side of New York, in the shadow of the Brooklyn Bridge, he never doubted his choice of career. At 14 he obtained a permit to walk around the unfinished steelwork of the Williamsburg Bridge. Steinman wrote poems about bridges, sent Christmas cards with sketches of his bridges, wrote books and articles containing technical studies on wind velocities and their effects on supension bridges and wrote popular layman's guides to bridges and their histories. Like the early bridgebuilders, who had to be great impresarios and promoters, Steinman, too, was a showman and sought fame and publicity. Perhaps this was a result of his difficulty in finding a job early in his career, but it tended to make his contemporaries look upon him as a publicity hound. Still, none could deny that Steinman was an energetic and competent engineer. His entry in *Who's Who* of 1950 was the third longest in the volume. By the time he was chosen to design the Henry Hudson he had been awarded the Columbia University Medal for Excellence. His adventurous career had included designing bridges throughout the United States, and in Canada, Australia, England, Germany, Spain, South America and Siam.

Steinman believed that "bridges are an index of civilization" and that the bridge designer of this era must be both engineer and artist. He called for bridge engineers to create "beauty and harmony of composition through beauty of line, form and proportion, through color and illumination."

The Henry Hudson Bridge won honorable mention among steel bridges costing more than $1 million completed in 1936. It was then the longest fixed steel arch in the world. It followed a practice of which Steinman was quite fond—painting his bridges in hues which would blend in with the natural landscape rather than using the traditional metallic gray. Steinman directed his Mount Hope Bridge in Rhode Island be painted a pale green to reflect the yacht-gliding green of Naragansett Bay. The Henry Hudson Bridge was originally painted deep green to blend in with the forest-like setting of Riverside and Inwood Parks.

Steinman also wrote that steel is a material capable of its own aesthetic, which "should not be regarded merely as a skeleton to be concealed or clothed in some foreign raiment." Accordingly the main towers of the Henry Hudson arch would be built of steel rather than the concrete traditional in arch bridges. To provide these 115-foot tall piers with the architecturally needed element of mass at the ends of the arch, he designed each tower with four bents similar to those on the arch but with 20-foot wide columns braced longitudinally. The outer and inner columns are braced in pairs with X-bracing; the inner columns are also connected at the deck by a semicircular arch portal and bracing in the form of the Cross of St. George.

During the construction of the northern approach to the Henry Hudson Bridge, a controversy erupted over an alleged Indian princess who taught pottery in the old woods of Inwood Park. In what Robert Moses called "a romantic escape from boredom," an outcry was raised by "parlor conservationists who had never climbed Inwood Hill" about removing the princess' working area and some famous old trees which surrounded it. A compromise was reached, and the authority took care to save the good trees which are left on the hill.

The bridge opened on December 21, in a wild rainstorm. The ceremony was held in the toll structure but it was halted when Mayor LaGuardia insisted upon listening to the broadcast of King Edward VIII renouncing the throne of England to marry a commoner and become the Duke of Windsor.

The Henry Hudson Bridge proved a financial success the first year it opened. The investors quickly retained Steinman to add the second deck and enlarge the approaches. At the completion of this second deck in 1937, a bronze 16-foot statue of Henry Hudson finally took its place on the Henry Hudson Memorial column in Riverdale, opposite 227th Street, north of the bridge. The original design for this statue had been created by Karl Bitter, a sculptor who died in 1915. The Henry Hudson Bridge Authority commissioned an old friend of the sculptor to reproduce the original design from a photograph of Bitter's model.

At the same time Moses was also in the midst of directing the completion of the Marine Parkway and Cross Bay Boulevard Bridges across Jamaica Bay to Rockaway peninsula. Here, vast stretches of undeveloped meadowland had changed little since only Indians lived in New York. The enormous bay contained by the Rockaway barrier beach was and is a haven for migratory fowl. According to Moses, Rockaway was a waterfront slum, which the authority and the parks department had cleared in conjunction with the building of the bridges. Some opposition was posed by conservatives who sentimentally fought to preserve a tawdry amusement area. However,

Opposite above: Internal view of Marine Parkway Bridge vertical lift span. In closed position this span is 50 feet above mean high water. It can be raised to 150 feet, as in lower photo. The three spans are equal in length. Each is 540 feet. It was designed by David Steinman.

Motorcade crosses Bronx-Whitestone Bridge as part of Opening Ceremonies on April 29, 1939. Notice the absence of a stiffening truss along the roadway.

COURTESY OF THE BROOKLYN PUBLIC LIBRARY BROOKLYN COLLECTION

Moses wanted the area to be a natural setting for recreation. The Cross Bay Bridge had been a rickety trestle bridge which was torn down when the Cross Bay double leaf bascule bridge and causeway over the bay replaced it.

The Marine Parkway Bridge, another Steinman creation, was completed on July 3, 1937, just in time for Independence Day traffic to reach the new Jacob Riis Park on Rockaway. It is an imposing lift bridge with tapered steel towers which have a scroll effect at the top. The design expresses the purpose of the counterweights instead of hiding them. The bridge is 4,022 feet long, and has three main spans with a 540-foot center lift span. This is the world's longest highway lift span, and the roadway was the first to use an iron grill. It is non-skid and easy on tires. It also automatically removes snow and wears well.

The bridge cost $16 million and took less than one year to build. On opening day, in the midst of the ceremonies, the lift span had to suddenly be lowered to permit a fire engine to get to a fire near Marine Park.

These two bridges, however, did not arouse as much excitement and appreciation as the new Bronx-Whitestone Bridge which was being built for the 1939 World's Fair. Moses was proud that this 2,300-foot suspension bridge, the world's fourth longest at the time, was financed entirely privately and was therefore free of government interference. Ammann, who followed the new trend toward slimmer bridges, designed an ethereal four lane span, eschewing the use of Warren trusses along the side of the deck in favor of low plate girders. The span was constructed in 23 months, opening 60 days ahead of schedule.

At the opening Moses described the new Bronx-Whitestone Bridge as "architecturally the finest suspension bridge of them all, without comparison in cleanliness and simplicity of design, in lightness and absence of pretentious ornamentation. Here, if anywhere," Moses added, "we have pure, functional architecture."

The *Times* commented that the new bridge's "freedom from heavy structural lines and ornamentation gives a breathtaking grace to the 2,300-foot center span." And Elizabeth Mock's *The Architecture of Bridges* described the Bronx-Whitestone Bridge as having a "fine spun elegance of outline and detail that was unique among modern suspension bridges." Certain reservations were held about the design of the two 377-foot towers, each of which contains a vast central arch and six small decorative arches as a border at the top. Six small arches were considered "obviously something of an affectation".

Aesthetics alone do not make a bridge successful. The Bronx-Whitestone Bridge suffered from a fault which stemmed from the fashion for slim, elegant bridges. It was soon noticed that the 11-foot high flanking girders that had been substituted for the usual deep Warren trusses responded unpredictably to exceptional wind pressures. Although suspension bridges have a certain "breathing room" and are expected to deflect vertically and horizontally to an extent, the Bronx-Whitestone Bridge produced alarmingly high vertical motion under some wind conditions. This motion was also observed across the country in Washington, where the Tacoma Narrows Bridge, designed by Leon Moisieff, had just opened.

The Tacoma Narrows Bridge also had four lanes. It was 500 feet longer than its New York cousin, but it was designed along very similar lines. There was also no truss along its deck. The length-width ratio in the Tacoma Narrows Bridge was 2,800:39; in the Bronx-Whitestone 2,300 : 74. The oscillations in the Tacoma Bridge were so obvious to everyone that it earned the nickname "Galloping Gertie," a name previously held by an early suspension bridge across the Ohio River. Rather than deterring users, the Tacoma bridge's oscillating roadway seemed to make the crossing all the more exciting.

During a heavy windstorm in November, 1940, four months after the Tacoma Narrows Bridge opened, its roadway began to buckle and twist like taffy and eventually ripped apart. The winds had caused extreme vertical oscillations which were magnified by the plate girders. This was one of the most famous bridge failures. A film of the wildly waving span taken by an alert newsman is included in the library of almost every engineering school. Studies of this film helped engineers develop a new science of bridge aerodynamics, which combined the deflection theory of suspension bridges, the mathematical theory of vibration analysis and the science of aerodynamics. The new theories permit engineers to predict wind effects on bridges. Besides the old method of increasing the stiffness of the structure to resist wind action, the sections of a bridge are now designed for aerodynamic stability, thereby eliminating the cause of oscillation.

Moses recalls that shortly after the Tacoma catastrophe, a lady in a small car driving towards Queens saw a lamp post on the bridge moving up and down. Remembering Tacoma, she became frightened. She jammed on the brakes, jumped out and began run-

It was soon discovered that high winds caused the slender span to shake alarmingly. Engineers fashioned a stiffening truss and added several radial stays to brace the bridge.

COURTESY OF THE TRIBOROUGH BRIDGE AND TUNNEL AUTHORITY

ning towards Whitestone. A second lady almost collided with the first car, leaped out, and followed. A man at the observation booth ran out to reassure them. The engineers and architects of the Bronx-Whitestone Bridge were soon called in. Ammann kept reassuring Commissioner Moses that this bridge was absolutely safe. "We kept replying 'that doesn't make a damn bit of difference if drivers won't use it,' " Mr. Moses remembers telling the great engineer.

Remedial steps were soon taken after studies of wind pressures were made. The first step was to install several diagonal stays, similar to those on the Brooklyn Bridge, from the tower tops down to the plate girders on the roadway. This method had been used previously by David Steinman when he realized that his Thousand Islands Bridge was not stable enough and suffered from similar oscillations. When he had heard of the problems on the Tacoma Narrows Bridge, he had offered the same solution before the bridge was destroyed, but the builders believed the measures unnecessary. Steinman felt the incident could have been prevented if his advice had been heeded.

In 1946 a 14-foot-high Warren truss was added over the plate girders of the Bronx-Whitestone Bridge. The sidewalks were eliminated, and the roadway lanes were slightly widened. According to Moses, this made it appear to travelers that the trusses were merely part of a widening operation. In effect, it constituted a kind of marketing psychology, preventing users and potential users from wondering about the safety of the bridge.

Many drivers who frequently use this bridge still assert that the Bronx-Whitestone is the most unstable bridge in the area. A violent storm in November 1968 which brought 70-mile-an-hour winds to the metropolitan area caused enough vertical oscillation on it to make motorists aware of the movement. Although the oscillations measured only 10 inches, it was enough to make the drivers feel quite uncomfortable. Robert Moses' contention is that a beautiful bridge is worthless if drivers don't feel safe on it.

It took two decades before the Triborough Bridge and Tunnel Authority embarked upon its next major bridge. In the meanwhile it had completed two tunnels, and had become heavily committed to a highway network for New York City. When it proposed the Throgs Neck Bridge to be built more or less parallel to the Bronx-Whitestone but some distance east, the purpose was to alleviate the great overcrowding on that bridge. Although the motor car was not yet seen by urban planners as the nuisance it now seems, and the earlier Triborough Bridge Authority bridges had been greeted with enthusiasm and public pride in public work, the proposed new bridge met unexpected opposition.

It has recently been stated by author-sociologist Jean Gottman in her book *Megalopolis* that Parkinson's Law applied to highway technology. Gottman suggested that as soon as a roadway is built, it attracts more users and becomes filled to capacity. More parkways invite more users, more cars, and overcrowding. Statistics bear out her theory. Although the local citizens did not have these statistics, they intuitively realized that the neighborhoods of Bayside, Queens and Throgs Neck in The Bronx would be destroyed by the invasion of workmen, noise and a new swath of highway cutting across the community. The prospect aroused resentment. Where once citizens had been delighted at the amount of work a structure like the bridge would provide, suburbia of the 1960's demanded more respect for its greenlawned communities.

Robert Moses felt residents had been "misinformed" about the approaches in Bayside, and he made correct information available directly to the people. The Authority opened an information and relocation office at the bridge site, staffed by real estate and engineering personnel and equipped with a scale model of the Throgs Neck Bridge and its approaches. Only 421 houses were involved in clearing space for the new right of way. The Authority offered, as a placating gesture, to relocate these homes on larger plots than those on which they were originally situated. To accomplish this, the Authority bought up the Bayside and Oakland golf courses. On The Bronx side of the 1,800-foot span was the N. Y. State Naval Academy, where a land fill project was exchanged for right of way for the new bridge.

Vehicular statistics for the Bronx-Whitestone and Throgs Neck bridges certainly bear out Gottman's theory. While the Bronx-Whitestone Bridge carried 33.2 million vehicles in 1960, by 1961, the opening year of the Throgs Neck Bridge, 40% of the traffic had been siphoned off. However, by 1966, usage of the Throgs Neck Bridge climbed to 30.6 million vehicles, approximately the same amount which was now flowing over the Whitestone again. Thus we have two bridges and twice the original traffic. This, of course, does not necessarily mean the demand would not have grown anyway.

The cost for the new six lane bridge reflected the inflationary trend of the times. The $92 million cost was almost triple the price tag of the Bronx-Whitestone Bridge. Again the cost was borne by private bondholders.

Recognition of the fact that grandiose public works may disturb the quality of life in many communities and that the never-ending progress paving the way for automobile traffic is not necessarily a boon to mankind, took root with those who objected to the Throgs Neck span. Opposition crystallized when it caused serious delays in the building of the greatest suspension bridge of them all — the Verrazano-Narrows Bridge.

COURTESY OF THE TRIBOROUGH BRIDGE AND TUNNEL AUTHORITY

The Throgs Neck Bridge, looking north from Clearview Park. Its main span is 1,800 feet long. Towers rise 360 feet.

COURTESY OF THE AUGUST ROCHLITZ COLLECTION

The old Beach Channel Bridge to Rockaway in 1925. Fishermen parked their cars right on the viaduct and the crossing was considered a bottleneck. The TBTA built the Cross Bay Bridge in 1938 to replace it.

COURTESY OF THE TRIBOROUGH BRIDGE AND TUNNEL AUTHORITY

View from Staten Island tower shows liner Queen Mary sailing under Narrows construction site.

VERRAZANO-NARROWS BRIDGE

...AND STILL CHAMPION

If cities still had protecting gods, as they did in Classical Times, New York's would be perched atop the elegant soaring portals of the Verrazano-Narrows Bridge. Once, immigrants, sailors and liner passengers were greeted by the welcoming arm of the Statue of Liberty. Today, seafarers associate New York harbor with the loftiest symbol of American technology: the world's longest suspension bridge.

Although New York's oldest suspension span—the Brooklyn Bridge—has long held first place among bridge enthusiasts, the Verrazano-Narrows Bridge, completed in 1964, has quickly secured second place. Its finely curved roadway and streamlined 70-story towers arch majestically over the entrance to New York. Its form is so simple, so successful, so serene, that nothing about the construction calls attention to itself except the absolute vastness of the span. The bridge, as Robert Moses noted on its opening day, is a "triumph of simplicity and restraint over exuberance." The bridge has a clear span of 4,260 feet—60 feet longer than the former record holder, the Golden Gate Bridge. Because of its great length, the curvature of the earth had to be taken into account when the towers were erected. Although both towers are exactly perpendicular to the surface of the earth, they are 1 5/8 inches farther apart at the top than at the base.

In the Verrazano-Narrows Bridge we have the supreme achievement of the modern bridgebuilder. The bridge took almost 75 years to gain acceptance by politicians, government agencies and citizens, but the end product is a sheer harmony of steel and concrete, the last great bridge from the drawing board of Othmar Ammann, who was born about the same time the Narrows crossing was first proposed.

Bridge and tunnel advocates who wanted to link Staten Island with the rest of New York had been unsuccessful for about 60 years when Robert Moses, chairman of the Triborough Bridge and Tunnel Authority, began campaigning for a bridge across the narrowest part of New York's hourglass-shaped harbor. The first advocate of the crossing had been the B. & O. Railroad in 1888. The railroad had previously built a swing bridge from New Jersey to Staten Island, but was still using ferries to carry its goods from Staten Island to Brooklyn and New York. To bring its freight directly into New York City, the B. & O. commissioned an engineer to design a tunnel from Staten Island to New York. This plan never got past the drawing board.

Another tunnel, from Brooklyn to Staten Island, was supported by New York's transit-minded Mayor John Hylan in 1923. Hylan, who advocated the general development of Staten Island, got $500,000 appropriated for a combination freight and passenger railroad tunnel. The project fizzled out, leaving two tunnel shafts, "Hylan's Holes," as mocking testimony to the history of such futile attempts. Residents of Bay Ridge in Brooklyn believed Robert Moses' attempts to bridge the Narrows would meet the same results.

In 1926, David Steinman, a cocky, passionate, energetic New York engineer, the country's youngest engineering professor, proposed the first bridge for the location. As Steinman's first love was the Brooklyn Bridge, he had modeled his career after that of John Roebling, taking an interest in philosophy, aesthetics, religion and political matters as well as several engineering fields. At Columbia University he had obtained three graduate degrees in engineering simultaneously. His doctoral thesis was widely read. His baptism in bridgebuilding came when he was called in to assist Gustav Lindenthal in building the Hell Gate Bridge. He replaced Ammann, who was drafted into the Swiss Army during World War I. Steinman's involvement in and love for his craft were so intense that he claimed to be personally acquainted with every rivet on Lindenthal's great railroad bridge. His enthusiasm was boundless, but it seems that his effusiveness and egotism put off many of his colleagues.

Steinman dreamed of building the world's longest suspension bridge to span the Narrows. Although several 3,000 foot spans to bridge the Hudson had been proposed by 1926, conclusive studies had yet to be made as to whether a suspension bridge of such length was feasible. The longest suspension bridge in the United States was a mere 1,700 feet long. Steinman's proposed Liberty Bridge was to be over 4,000 feet long. If the engineering aspect of Steinman's Liberty Bridge was daring, his aesthetic conception of the towers—the most prominent feature of any suspension bridge—was overwhelming. His sketches reveal a frilly steel frame design consisting of tiers of Gothic tracery arches topped by an ornate steel spire that would house observation balconies and a carillon of bells that "will peal out anthems of Liberty to those who enter through this Nation's gateway; and a beacon light that will send out its beams as a symbol of Liberty to guide, welcome and

COURTESY OF THE NEW YORK MUNICIPAL REFERENCE LIBRARY

David Steinman proposed this "Liberty Bridge" for the Narrows in 1926 as a tribute to World War I casualties.

inspire those who have crossed the seas to come to our shores."

Steinman had learned the lessons of John Roebling and Gustav Lindenthal, and was wary of politicians. An adherent of conservative economics, Steinman undertook to convince wealthy entrepeneurs that his Liberty Bridge would be a gold mine. These New Yorkers, buoyed by the smoothly running boom of the '20s, and impressed by Steinman's credentials, formed the Interborough Bridge Company to finance the bridge as a completely private business venture. Still, as it was necessary to get a government charter, a Staten Island Congressman was collared to introduce the bill in the House of Representatives. The bill went before Congress the day before it adjourned in 1926. Opposition came from a lone Congressman who roared: "I don't want private capital to profit at the expense of the people." This fiery dissenter was Fiorello LaGuardia. Congress adjourned and the charter was never granted.

COURTESY OF THE NEW YORK MUNICIPAL REFERENCE LIBRARY

Liberty Bridge's elaborate 800-foot steel towers were to contain observation balconies, carillons and beacon lights.

LaGuardia was not at all opposed to the bridge itself. During the Depression, when he became mayor of New York, he advocated the Liberty Bridge as potentially a great public work to relieve unemployment at Federal expense. The proposed Narrows crossing, under the LaGuardia Administration, became an integral part of the master plan of express highways, parkways and major streets aimed at relieving Manhattan's traffic congestion. The arteries of this plan had been plotted by city prime mover Robert Moses.

Moses carefully steered clear of designating the crossing as either a bridge or a tunnel. He was on the outs with the Roosevelt Administration, and realized he had to leave all options open to secure the permission of the State, Federal highway funds and the approval of the Navy. After winning a prolonged battle to get Federal financing of Triborough Bridge while he maintained his "unwanted" leadership of the Triborough Bridge Authority, Moses had proceeded to build the Bronx-Whitestone, Marine Parkway, Henry Hudson and Cross Bay Parkway Bridges with neither Federal interference nor Federal funds. However, when it came to a new Battery crossing between Brooklyn and Manhattan to relieve downtown congestion, Moses again had had to tangle with the Roosevelt Administration.

The conflict started when Eleanor Roosevelt, in her syndicated column, *My Day,* complained that the proposed bridge would mar the beauty of the New York skyline. In addition, the Department of the Navy turned down the bridge plan, claiming that enemy planes might someday destroy the bridge and thereby blockade the Brooklyn Navy Yard from the sea. (The Brooklyn Navy Yard was already located up river from the Brooklyn and Manhattan Bridges, and the *Daily News* suggested that the Department of the Navy was deferring either to Roosevelt, or to the desire of various military officers who fancied a game of polo on Governor's Island, and thought the shadow of the bridge might ruin their game.) It is more than possible, as the *Daily News* noted at the time, that the attack was an instrument of Franklin Roosevelt's vindictiveness toward Moses.

Moses disgustedly changed his plans for the Battery crossing to a tunnel, maintaining that in his opinion a tunnel was a "tiled, ventilated, vehicular bathroom, smelling faintly of monoxide and inviting claustrophobia."

So, losing the battle for a Battery Bridge in the

early '40s, Moses turned his energies towards promoting the Narrows crossing. Taking into account the opposition he would have to confront, Moses predicted with patient irony that it would take 20 years from the time he began his campaign to the time the bridge would be built. He realized it would take years of intermittent skirmishes with Federal power wielders and local clamorers to achieve the results he wanted, but Moses also saw the inevitable power of progress on his side and was willing to wait.

First, he had plans for a tunnel drawn up. The designers proved, to his delight, that a two lane vehicular tunnel run under bedrock in the Narrows would cost more than a two-level 12 lane suspension bridge. Once that was proved, Moses began promoting the bridge.

Publicity during the late '40s and early '50s seems to ascribe the design for the Narrows bridge to David Steinman. Although Steinman had been contemplating retirement, the imminent building of the bridge of his dreams brought him back into very active practice. The bridge was still referred to in many newspaper articles as the Liberty Bridge, and now Steinman was on a one-man campaign to beautify bridges by means of multi-colored illuminations. "Painting with light will be an integral part of its design," Steinman said, setting forth plans to bathe the structure in rainbow night lights.

However, Robert Moses, who had worked with both David Steinman and Othmar Ammann on various bridges around New York, was more in favor of Ammann's austere designs than of Steinman's flamboyant, almost baroque concoctions. When he sensed that political opinion was beginning to favor the new Narrows bridge, Moses asked Ammann, who had quit the Port Authority and had started a private consulting firm, to start submitting sketches.

While the engineers were busy creating paper masterpieces, Moses was fighting gut battles with the War Department to get permission to build the bridge. The War Department kept picturing the catastrophic effect a bombing raid might have on the harbor—a fallen snarl of steel and cables would plug up the bay. Five years were spent debating this contingency. Finally, engineers convinced the defense establishment that the only way the giant bridge could be destroyed would be by a direct hit by an atomic bomb, which would eliminate not only the bridge but also the Navy Yard. They also demonstrated that even if the bridge were in fact to be demolished (over 1600 tons of bombs were dropped on Scotland's Firth of Forth Bridge during World War II and it was not destroyed) a passage could be cut through the debris within 36 hours. In addition, Ammann's design had a center span 237 feet above high water level with a full load of traffic at a temperature of 120 degrees (lower temperatures made the suspension span higher). This left clearance for the Queen Mary, the tallest masted ship then afloat.

Another six years elapsed before Moses completely convinced the State Legislature and City Council to go ahead with the last enabling acts to permit the Port Authority and Triborough Bridge and Tunnel Authority to go ahead with the span. Moses felt the TBTA couldn't afford to build the bridge alone and convinced the Port Authority to supply some preliminary funds. However, by 1959 the TBTA had come up with funds and the Port Authority stepped out of the picture after participating only in the preliminary soundings for the towers.

COURTESY OF THE TRIBOROUGH BRIDGE AND TUNNEL AUTHORITY

Aerial view of completed crossing shows part of Brooklyn's Fort Hamilton in the foreground.

COURTESY OF THE TRIBOROUGH BRIDGE AND TUNNEL AUTHORITY

COURTESY OF THE TRIBOROUGH BRIDGE AND TUNNEL AUTHORITY

Top: The foundation caisson of the Brooklyn tower contained 66 17-foot diameter circular wells. Dredging started in 1961. Above: Fifth tier of the Brooklyn tower was put in place in the summer of 1962. Completed towers are 623 feet high.

Moses scored another victory in 1957 when the Army let him use land at the cores of the Fort Hamilton and Fort Wadsworth, military reservations on either side of the Narrows. In return the TBTA promised to spend $24 million building new and more efficient Army facilities at the two bases.

After dealing with the Department of Defense and New York politicians, Moses had to tilt with the local citizenry, which had formed a "Save Bay Ridge" movement and was lobbying in the State Legislature to defeat the bridge. Opposition was strongest in the tidy, Italian middle-class community of Bay Ridge, threatened not only by the bridge but also by the 12-lane approaches, the so-called "Colossus of Roads," which would dislocate close to 8,000 Brooklyn residents. In Staten Island a certain grudge was held against the bridge by those who wanted to preserve the rural quality of the island, but they were heavily outweighed by business and real estate interests which thought they could make a killing in the development of the island.

The clamor of the anti-bridge contingent was compounded by intense criticism from such luminaries as urbanologist Lewis Mumford. Mumford had profound praise for the Brooklyn Bridge in the 1920's, but he found the proposed Narrows Bridge "a menacing error . . . a terrifying project." He pointed out that the project would cost more than the Panama Canal and was being built only because "money is difficult for the hungry local sharks and minnows to resist." The opposition to the bridge was so active, vocal and desperate that it convinced the State Legislature, acting in "the bliss of ignorance" according to the pro-bridge *New York Times,* to defeat Moses' Bay Ridge approach proposals twice. Both times Governor Averill Harriman vetoed the legislative acts at Moses' request.

Meanwhile, David Steinman, seeing an opportunity to get back into the competition for the bridge contract, independently submitted a design for a bridge that bypassed Bay Ridge and Staten Island completely and linked North Brooklyn with Bayonne. Moses countered Steinman's scheme in a letter to Mayor Wagner, authoritatively writing it off as chimerical, and a financial impossibility. But new impetus was given to the anti-bridge faction when Nelson Rockefeller ran for governor in 1958, with a campaign pledge to relocate the Brooklyn approach to the bridge. Finally, with what seemed to his opponents like characteristic chicanery, Rockefeller allowed Moses to convince him that to change approaches would cost an extra $50 million, making the Narrows Bridge project so expensive that it would be "beyond revival in this generation." So despite their cries and plaints and shouts and despite an 11th hour "Save Bay Ridge" battle in the City Council, ground was broken for the new bridge in September 1959. Moses grandly noted that although local opinions must be heard, it was the bridge that reflected the "genuine interest of the nation."

As a small consolation to the Italian community, the State Legislature complied with the request of Brooklyn's Italian Historical Society to name the new bridge the Verrazano-Narrows Bridge after Giovanni da Verrazano, the Florentine explorer and mapmaker who had sailed into New York harbor in 1524, claiming everything in sight for the King of France. There had been some opposition to naming the bridge for Verrazano. At the groundbreaking ceremonies for the bridge the Staten Island Chamber of Commerce had hired a plane to circle over the proceedings with a banner bearing the advice, "Call it the Staten Island Bridge." On the other hand, the ferry boat which took dignitaries from downtown Manhattan to the Fort Wadsworth ceremonies, bore the name *Verrazano.* Robert Moses had it known that he favored the Staten Island designation, but relented and didn't challenge the State Legislature this time around. Another rather silly debate ensued on whether to spell Verrazano with one z or two. The explorer seems to have spelled it both ways. The TBTA insisted that Verrazzano should have two z's while the head of the Italian Historical Society was a one z man. A member of the society was promptly dispatched to Florence, and returned with evidence that one z was the more customary spelling.

This tribute to an Italian explorer hardly lightened the suffering of those whose homes were demolished for the bridge approach. In his book *The Bridge,* Gay Talese recounts the anguish of these dislocated Bay Bridge residents, but Talese also presents evidence of a new vitality in the adjoining communities, due not only to the money flowing into the area but also the noisy invasion of "boomers," experienced bridge men who eagerly migrated to New York to work on the world's longest, largest suspension bridge.

The foundations to support the 693-foot tall towers are steel and concrete open dredge caissons. Each giant concrete waffle, almost as large as a football field, has 66 circular wells, each well 17 feet in diameter. The massive caissons were sunk through manmade sand islands near the Brooklyn and Staten Island shores. Sand, gunk and mud spewed out of the wells until the caissons were sunk to a predetermined depth of 105 feet below high water on the Staten Island side and 170 feet on the Brooklyn side. It was 26 months before the caissons were firmly embedded in sand and clay. Finally, atop the caissons the pedestals appeared, the first visible signs of the bridge, rising 33 feet above water level to support the slender steel towers.

These graceful, sky-spearing portals look deceptively simple. But within each rivet-covered monolith

COURTESY OF THE TRIBOROUGH BRIDGE AND TUNNEL AUTHORITY

there is a honeycomb of 10,000 prefabricated steel cells interconnected by two elevators and more than 16 miles of ladders. Each cell is numbered, but the system proved such a labyrinth that many workers got lost. As a result, they were required to carry blueprints and miner's lights and there was an elaborate check-in and check-out system for each shift to make sure all workers were accounted for. Each tower weighs 27,000 tons and the bridge altogether contains three times as much steel as the Empire State Building.

The wedge-shaped concrete anchorages, far more massive than most Egyptian pyramids, are taller than 10-story buildings. Together they contain 375,000 cubic yards of concrete. The base of the Staten Island anchorage lies 76 feet below surface. The Brooklyn anchorage extends 52 feet below the ground. The base of each anchorage is 230 feet wide and 345 feet long, roughly the area needed to place two football fields side by side.

The next big job after the anchorages and towers was the weaving of the four 36-inch thick cables from which the double roadway is suspended. Each cable contains 26,108 strands of galvanized steel wire the thickness of a pencil. Altogether, 142,500 miles of wire—39,000 tons of it—were used. The cables alone cost more than the entire Golden Gate Bridge.

The project entailed so much work that three steel companies were contracted to do the job. Bethlehem Steel built the Staten Island tower and Harris Structural Steel Co. built the Brooklyn tower, each following identical specifications. U. S. Steel's American Bridge Division got the biggest order—everything else, including the cables and roadways.

American Bridge went about its giant chore by leasing a 22-acre assembly yard in Jersey City. While the spinning wheels shuttled to and fro weaving the cables, the company's ironworkers assembled, one by one, 71 sections of roadway, each weighing 388 tons. These were carried to the site by barge, then lifted into place with delicate precision. The job was done by lifting struts which raised each steel section into position to engage the vertical suspender hung from the main cables. With the addition of each road section the cables took on a new contour. The first section alone pulled them down 20 inches. At each instant the exact stresses and strains acting on every major component had to be calculated. A battery of computers handled this part of the job and provided reassurance that the roadway sections, which often seemed to be askew, would come into exact alignment when the loading was complete. When the entire two-level floor had been put in place the cable arc had been lowered nearly 28 feet. Even then, the curve of the cables was kept shallow to heighten the sense of sweep.

In a statement for *Life Magazine,* before the completion of the bridge, designer Ammann modestly pronounced himself pleased with the results. "The design of the bridge is largely determined by the conditions given—topography, geology, navigational requirements of the area. But within limitations," he explained, "the designer can still express himself. It is an attractive bridge," the Swiss engineer conceded. "It represents certain engineering advances."

Construction was halted several times because of high winds, hurricanes and other foul weather conditions. Once, after a 19-year-old ironworker plunged to his death from the catwalk during cable spinning, a four day walk off ensued. The boomers demanded nets under the bridge. As it was impossible to lift components through netting, the workers had to compromise and small nets were slung under the most precarious positions.

The roadway skeleton was finished by January 1964. By this time most of the restless boomers had moved on to the next great challenge and for the rest of year laborers filled in the gaps in the roadway with smaller steel beams and struts. Then they paved, did electrical work and painted.

On November 21, 1964, New York opened its 66th bridge over a navigable waterway, and perhaps marked the end of its golden period of bridgebuilding. Progress of bridge construction work has been speeded about six fold since the Brooklyn Bridge if one takes into consideration the magnitude of the project. After five years of work, the politicians reentered the scene to take credit for another splendid public work. They arrived, along with generals, admirals, business leaders, women in mink and pretty girls, in 52 stately black limousines which jammed up traffic for 4½ miles behind the site of the ribbon cutting ceremony.

There was no sign of the original Bay Ridge dissidents at the opening, but a new group took their place. On the Brooklyn shore, picket lines of teenagers protested the bridge's lack of pedestrian walkways. "Are feet obsolete?" one sign asked. In answer the TBTA said that surveys had shown that such a walkway would not only get very little use, and therefore not be worth its cost, but might also attract would-be suicides.

Another group of critics was the ironworkers themselves, who boycotted the ceremony. Responding to a call by their union leader, who denounced Robert Moses for not having invited the men "who put that bridge together piece by piece, strand by strand," they attended a memorial ceremony for the men who lost their lives on the bridge.

In the chilly November air, the speeches grew shorter. Robert Moses paid tribute to Othmar Ammann, who now at the age of 85 years was seeing his last great bridge. But Moses forgot to mention the Swiss engineer by name. When a stranger in the cold bleachers asked him how he felt on that day, however, Ammann, a bit startled, remarked, "Same as any other day."

Opposite: The base of the Staten Island anchorage lies 76 feet below ground. The 26,108 strands in each cable are embedded in cement.

Right: In early autumn 1963 boomers started placement of the cable bands. Special machine used in this process is called a "Kelly Wagon."

COURTESY OF THE TRIBOROUGH BRIDGE AND TUNNEL AUTHORITY

Kosciuszko Bridge, a high-level fixed span, was one of the last built by the city. It leaves Newtown Creek completely free to river traffic. There are proposals to replace many movable bridges with high-level structures, as they require no bridge tender and have no mechanisms to break down. Also they are less conducive to traffic jams. The bridge is named after General Tadeusz Kosciuszko, a Polish hero who served in the American Army during the Revolution.

SOME SMALLER BRIDGES

Opening day of the Pulaski bascule bridge over Newtown Creek in 1954. COURTESY OF THE BROOKLYN PUBLIC LIBRARY

COURTESY OF THE NEW YORK TRANSPORTATION ADMINISTRATION

Above: The Welfare Island Bridge permits traffic to go directly from Queens to the East River island (now Roosevelt Is.). When the lift span was finished in 1955, trolley service over the Queensboro Bridge to the island ceased.

Below: Aerial view of The Bronx's Eastchester Creek shows the Eastchester Bay Bridge in foreground, a railroad bridge center, and the Hutchinson River Pkwy. bascule bridge in rear.

COURTESY OF THE NEW YORK TRANSPORTATION ADMINISTRATION

COURTESY OF THE BROOKLYN PUBLIC LIBRARY

Above: Carroll Street Bridge, foreground, is over Brooklyn's Gowanus Canal. It is a 17-foot wide retractile bridge and is 3.5 feet above water when closed. The bridge in the rear is the Third Street Bridge. Right: Further along the canal is this bridge over a bridge. The Independent Subway constructed a high level span right over the Ninth St. bascule bridge.

PHOTO BY DAVID SPISELMAN

COURTESY OF THE NEW YORK TRANSPORTATION ADMINISTRATION

Inspection platform, or traveler, slowly glides under roadway of Williamsburg Bridge.

BRIDGE MAINTENANCE

Bridges, like living organisms, need room to breathe and stretch. They are subject to stress and chafing from tension and compression members. Traffic creates perpetual vibrations. The span contracts and expands with temperature changes. An auto accident or a stopped subway train in the center of a bridge can cause rivets to pop. Cement cracks in the anchorages. There is chafing in the suspenders, wear on girder flanges, corrosion under the roadways. "You have to live with a bridge to know what's wrong. Take an interest. It's a before and after kind of thing," explains city bridge engineer Edward Rubin.

So, undaunted by weather, more than 100 city carpenters, engineers, painters, riveters and steelmen daily poke into recesses and inspect surfaces on New York City's bridges to see what is wearing and what needs repair. They leap over railings, below roadways, performing bridge acrobatics to the delight and sometimes dismay of onlookers, who occasionally take them for suicides and call the police. Climbing hand over hand up the lacework of the towers, they test for loose rivets. They shrug off the danger of climbing up cables, which are equipped with railings for this purpose. "Fact is, coming down is more difficult than going up the cables," admits Martin Terry, who is supervisor for the crew on the Williamsburg and Kosciuszko bridges. Terry says he scrupulously observes all safety precautions despite the apparent nonchalance of the men in his crew. "The thing is, coming down, you have less control because gravity is affecting you and you may start gaining momentum without noticing."

Another inspection and maintenance duty consists of riding below the bridge's roadway in an unenclosed metal cage which creakily glides in slow motion beneath the span. It gives an unimpeded view of the stringers under the roadway, which get weakened by soot and corroded by salt strewn on the roadway on icy winter days.

Twisting around to get a better view of hard-to-see rivets and connections, the bridge crews seem excessively daring. But they cling to a mean between being too careful and too outrageous. "If you're too careful you start getting jittery," said a subway employee who walks the tracks over the Manhattan Bridge.

Steelmen inspecting the bridge tap each rivet with hammers and the ring of it determines whether they are tight or loose. The examiner makes a note and when there are enough corrections in one area, a steelman climbs up to tighten or replace them. Other chores include keeping the mercury-vapor lights clean and replacing them when they go out, as well as reporting the potholes that break up the roadway. Some of the workers allege that avoiding swerving, speeding autos is the most consistent danger a bridgeman encounters. "They get up there where there are no stoplights and they think they're on a racetrack," says Wilton Perry, who is chief of the city bridge painting staff.

Painting bridges is an important precautionary measure, and Perry, who has seen the city bridge painting staff decline by attrition for lack of funds in the last 30 years, remembers when each bridge had 20 or 30 men painting sections continually. Now the Department of Transportation has also made the bridge section responsible for highway overpasses and uses private contractors to paint sections of the major city bridges. This undercuts the efforts of regular city bridge painters to inspect the spans while they paint.

Painting the spans keeps down rust, and rust leads to serious corrosion of steel. The Port Authority considers it most economical to paint a bridge every seven years, although they admit that they could get along without painting a bridge like the George Washington for up to 12 years. However, chipping off the rust that accumulates in the ensuing five years costs more than the frequent paint jobs.

According to the Port Authority it takes 24,000 gallons of paint and wears out some 1,500 brushes to paint the George Washington Bridge, which has six million square feet of area requiring painting. The Triborough Bridge and Tunnel Authority reports that it uses 36,250 gallons for the Verrazano-Narrows Bridge. During World War II there was a great scarcity of aluminum paint and some of the bridges were neglected.

The Port Authority, unlike the city, does not retain a sizable number of bridge painters at all times. When it decides to paint one of its four spans it hires quite a number of specialists for the job. Although most are former seamen and some are even circus acrobats, a surprisingly large number of men turn up for the bridge painting jobs and not all of them are qualified. They are given three tests, one of which involves climbing to the top of the west tower of the George Washington Bridge, 640 feet above water, and walking along a beam eight feet wide and 20 feet long. "A man who isn't at home on this bridge painters' plaza is regarded as a poor risk," says the bridge painter foreman for the Port Authority.

A novel experiment aimed at lubricating the ungalvanized steel wires of the Williamsburg Bridge several years ago called for the steelmen to climb the cables of the Williamsburg Bridge, unwrap the cables and pour in fish oil, letting gravity take its course. The results indicated that the oil could not perform the function for which it was intended and it was best just to keep the cables tightly closed and well-painted.

Stopping vandalism and cleaning up debris is another part of the job for bridge maintenance crews. The location of the bridge determines the size of this problem, with the safest areas being industrial sections. Residential areas, according to the Department of Transportation, are sources of vandalism, with mischievous kids displaying delinquent behavior whether the bridges are in "good" or "bad" neighborhoods.

Kids are fascinated by bridges, and have been found climbing along the cables or under the roadways. The Manhattan Bridge's walkway had to be closed because students in a neighboring high school used to spend their lunch hours loitering on the bridge. The pedestrian path is on the span's outermost lane, and it is considered a temptation for suicides, as well as a perilous place for horseplay, according to the Department of Transportation. On the subway side of the walkway, a person can walk right onto the third rail. So, rather than risk lives, the path has been closed for nearly two decades. Similar measures have been taken recently on other bridges.

The city's bridge keepers are divided into three sections. One maintains the East River spans; another staffs the bridges in Brooklyn, Queens and Richmond; while a third watches over the spans on the Harlem River as well as those in The Bronx and Northern Manhattan.

Each movable bridge is manned around the clock by an operator and an attendant, so there is someone virtually living on the bridge at all times. In addition, inspectors visit these bridges on a rotating basis.

They are required to keep the bridges free and clear so that the span may be open to navigation at any time. Boats have the right of way over land traffic.

According to a former ironworker for the Department of Highways, many of the movable bridges are so idiosyncratic due to age, damage from passing ships and rust and rot at the water line, that they are very difficult to operate. "None of them work the way they are supposed to work. Many are 75 or more years old. The operators paint markers on the walls. The bridge's wheel goes around on a track, but the bridge doesn't open where the wheel indicates, but rather when it passes the wall marker."

Bridge keepers must also open and close the gates to stop traffic where this operation is not electrified; keep sidewalks, roadways and bridgehouses clean during summer and winter; patrol the bridge; make out accident reports in the case of unusual occurrences; keep all lights clean and in shipshape condition; flag navigation where bridges cannot be opened immediately; keep the operator bell signal for operation of bridges and keep records on tally sheets of boats sailing through the draw.

Because they have to keep the bridges and roadways clean, a permit is made out each year to the men in charge of each bridge, allowing them to tap the hydrant nearest the bridge and spray the roadway. The bridge must be sprayed daily in very hot weather to prevent expansion.

Temperature changes can alter the spans very seriously. In the case of the Metropolitan Avenue Bridge in Queens, the roadway expanded to such an extent after a week of hot weather, that the bridge's leading edge (or metal closure) wouldn't close properly. Bridge workers had to be sent down with acetylene torches to burn off these edges. When winter came they were again sent there to weld on a new edge in order for the roadway to close.

Such emergencies are not unusual for the men who conscientiously attend the city's spans. One of the city's staff engineers described the backlog of problems on the city's bridges as "emergencies on top of emergencies," with the bridge staff often powerless to take any measures. Now that they are under the jurisdiction of the Department of Transportation, the bridge crews and an independent crew of inspectors are only allowed to "suggest priorities." "What is a priority?" questioned a staff engineer. "We can't tell them that a bridge is in immediate danger of falling down. Nobody can say that. They always want to know if it will fall like the West Side Highway. But that's very hard to say." Meanwhile, he shook his head in disapproval, "You can't take care of a bridge like you take care of a house."

COURTESY OF THE NEW YORK TRANSPORTATION ADMINISTRATION

Each East River Bridge has its own maintenance crew. Here workman replaces one of Brooklyn Bridge's mercury vapor lamps.

BRIDGES UNDER THE JURISDICTION OF THE

NAME	WATER CROSSING	LOCATION
BROOKLYN	East River	Park Row near Frankfort St. Adams near Tillary St.
MANHATTAN	East River	Canal St. and Bowery Nassau St. and Flatbush Ave. Ext
WILLIAMSBURG	East River	Clinton & Delancey Sts. Broadway & Roebling St.
QUEENSBORO	East River	2nd Ave. and 59th St. Crescent St. & Bridge Plaza
WELFARE ISLAND	East River (E. Channel)	Welfare Island Vernon Blvd. and 36th Ave.
WILLIS AVE.	Harlem River	First Ave., 125th St. Willis Ave., 134th St., Bruckner Blvd.
THIRD AVE.	Harlem River	Lexington & Third Ave., 129th St. Third Ave., 135th St. & Bruckner Blvd.
MADISON AVE.	Harlem River	Madison Ave. & 136th St. East 138th St.
145th STREET	Harlem River	West 145th St. East 149th St.
MACOMBS DAM	Harlem River	West 155th St. Jerome Ave.
WASHINGTON	Harlem River	West 181st St. University Ave.
UNIVERSITY HEIGHTS	Harlem River	West 207th St. West Fordham Road
SHIP CANAL	Harlem River	Broadway near 225th St.
WARDS ISLAND PEDESTRIAN BR.	East River	E. 103rd St., Wards Island
174th STREET	Bronx River	East 174th St.
EASTERN BLVD. (Ludlow Ave.)	Bronx River	Bruckner Boulevard, Edgewater Road
WESTCHESTER AVE.	Bronx River	Westchester Ave.
UNIONPORT	Westchester Creek	Bruckner Boulevard, Zeraga Ave.
PELHAM	Eastchester Bay	Bruckner Boulevard Pelham Bridge Road
CITY ISLAND	Pelham Bay Narrows	City Island Road City Island Ave.
EASTCHESTER	Eastchester Creek	Boston Road & 233rd St.
HUTCHINSON RIVER PKWAY. EXT.	Eastchester Creek	Hutchinson River P'kway Ext.
HAMILTON AVE.	Gowanus Canal	Hamilton Ave.
NINTH STREET	Gowanus Canal	Ninth Street
THIRD STREET	Gowanus Canal	Third Street
CARROLL STREET	Gowanus Canal	Carroll Street
UNION STREET	Gowanus Canal	Union Street
METROPOLITAN AVE.	English Kills	Metropolitan Ave.
STILLWELL AVE.	Coney Island Creek	Stillwell Ave.
CROPSEY (Harway) AVE.	Coney Island Creek	Cropsey Ave.
THIRD AVE.	5th Street Basin (Gowanus Canal)	Third Ave.
MILL BASIN (h)	Mill Basin	Shore Parkway
OCEAN AVE. (b)	Sheepshead Bay	East 19th Street
PULASKI	Newtown Creek	Oakland & Eagle Sts. 11th St. & Jackson Ave.

NEW YORK DEPARTMENT OF TRANSPORTATION

TYPE OF BRIDGE	LENGTH INCLUDING APPROACHES IN FEET	MAX. SPAN IN FEET	CLEAR HEIGHT ABOVE M.H.W. IN FEET	TRAFFIC FACILITIES Roadway	Footwalks	WIDTH OF CHANNEL OPENINGS IN FEET	DATE OPENED TO TRAFFIC
Suspension	6,775	1.595.5	133	2-30'-0"	1-15'-7"	1,508.5	May 24, 1883
Suspension	6,855	1,470	135	1-35'-0" 1-23'-6" E. 1-22'-6" W.	1-10'-0"	1,230.25	Dec. 31, 1909
Suspension	7,308	1,600	135	2-18'-2" 2-19'-11"	1-17'-8"	1,517	Dec. 19, 1903
Cantilever	11,110 U.R. 7,449 L.R.	1,182	135	1-50'-10" 2-22'-10"	1-9'-9"	938 793	March 30, 1909
Lift	2,877.3	418	High 103 Low 40	1-34'-0"	1-6'-0"	403'-8½"	May 18, 1955
Swing	3,212.5	304	25.1	1-42'-0"	2-9'-0"	108	Aug. 22, 1901
Swing	2,800	300	25.8	2-26'-0"	2-9'-0"	100	Aug. 1, 1898
Swing	1,892	300	25	2-27'-0"	2-9'-0"	104	July 18, 1910
Swing	1,603	300	25.2	2-27'-0"	2-9'-0"	104	Aug. 24, 1905
Swing	2,540	408.5	29.2	1-40'-0"	2-9'-8¾"	165	May 1, 1895
Steel Arch	2,375	508.8	133.5	2-33'-0"	2-6'-3"	400	Dec., 1888
Swing	1,566 470 Ramp	264.5	25	1-33'-6" 1-28' Ramp	2-5'-8"	100	Jan. 8, 1908
Swing	557	265.5	24.7	1-34'-10"	2-5'-4½"	100	June 17, 1905
Lift	1,247.02	312.19	High 135 Low 55		1-12'-0"	300	May 18, 1951
Fixed	605	190	30.5	1-42'-0"	2-10'-6"	100	June 15, 1928
Bascule	634.19	118.69	26.6	2-34' 2-22'	2-7'-6"	70	Oct. 1, 1931
Bascule	363	83	17.2	1-60'-0"	2-7'-0"	60	Nov. 1, 1938
Bascule	601	75	17.5	2-58'-0"	2-7'-6"	60	(m) Aug. 1, 1918
Bascule	891	80	17.5	1-40'-0"	1-7'-6"	60	Oct. 15, 1908
Swing	1,470	164.75	12.3	1-32'-7"	2-6'-4½"	54.3	July 4, 1901
Bascule	292	127.67	12.5	1-53'-0"	2-8'-6"	75	April 8, 1922
Bascule	670.2	160.5	35	2-36'-0"	1-8'-0"	130.5	Oct. 11, 1941
Bascule	546	66.32	19	2-42'-0"	2-8'-0"	47	Aug. 27, 1942
Bascule	109	56	7.3	1-35'-0"	2-6'-0"	45.2	July 13, 1905
Bascule	360	56	7.3	1-30'-0"	2-8'-0"	40.4	March 31, 1905
Retractile	165	45.2	3.5	1-17'-0"	2-3'-7"	36.0	1889
Bascule	109	56	8.3	1-35'-0"	2-6'-0"	43.5	March 4, 1905
Bascule	440	111	10.72	1-53'-0"	2-5'-8"	86	March 27, 1933
Swing	264.3	250	5.7	1-21'-9"	2-7'-9"	40	R'dway Opened Jan. 8, 1929
Bascule	382.7	155	11.25	2-36'-0"	2-7'-4"	75	Dec. 20, 1931
Steel Girder	60	40.3	13	1-42'-0"	2-13'-6"	36.3	1889
Bascule	864.5	165	35	2-34'-0"	2-6'-0"	131	June 29, 1940
Fixed	470	46.2	7.8		1-8'-0"	39	June 9, 1917
Bascule	2,810	176.67	46.33	2-34'-0"	1-8'-0"	130	Sept. 10, 1954

BRIDGES UNDER THE JURISDICTION OF THE

NAME		WATER CROSSING	LOCATION
GREENPOINT AVE.		Newtown Creek	Greenpoint Ave.
GRAND ST.		Newtown Creek	Grand Street Grand Ave.
KOSCIUSZKO (Meeker Ave.)		Newtown Creek	Morgan Ave. & Meeker Ave. Borden Ave. & Laurel Hill Blvd.
BORDEN AVE.		Dutch Kills	Borden Ave.
MIDTOWN HIGHWAY CROSSING		Dutch Kills	Midtown Highway
HUNTERS POINT AVE.		Dutch Kills	Hunters Point Ave.
FLUSHING		Flushing River	Northern Blvd.
ROOSEVELT AVE.		Flushing River	Roosevelt Ave.
WHITESTONE PARKWAY		Flushing River	Whitestone Parkway
LITTLE NECK		Alley Creek	Northern Boulevard
HOOK CREEK		Hook Creek	Rockaway Boulevard Rockaway Turnpike, Nassau County
NORTH CHANNEL		North Channel	Cross Bay Boulevard
HAWTREE BASIN		Hawtree Basin	Near Nolins Ave.
LEMON CREEK		Lemon Creek	Bayview Ave.
FRESH KILLS		Richmond Creek	Richmond Ave.

BRIDGES UNDER THE JURISDICTION OF THE

NAME		WATER CROSSING	LOCATION
TRIBOROUGH (3 Spans)	1.	East River	Grand Central Parkway
	2.	Bronx Kills	Bruckner Expressway
	3.	Harlem River	125th St.
HENRY HUDSON		Harlem River at Spuyten Duyvil	Henry Hudson Pkwy.
MARINE PARKWAY		Rockaway Inlet	Flatbush Ave. to Marine Pkwy.
CROSS BAY PARKWAY		Jamaica Bay	Cross Bay Blvd.
BRONX-WHITESTONE		East River	Cross-Island Pkwy.-Hutchinson Riv. Pkwy.
THROGS NECK		East River	Clearview Expwy.-Cross Bronx Expwy.
VERRAZANO-NARROWS		The Narrows	Bklyn-Queens Expwy.-Staten Is. Expwy.

BRIDGES UNDER THE JURISDICTION

NAME	WATER CROSSING	LOCATION
GOETHALS	Arthur Kill	Staten Island Expwy.
OUTERBRIDGE CROSSING	Arthur Kill	Richmond Pkwy.
GEORGE WASHINGTON	Hudson	178th St. & Ft. Lee
BAYONNE	Kill Van Kull	Willowbrook Expwy.-Kennedy Blvd.

RAILROAD

NAME	WATER CROSSING	LOCATION
HELL GATE	East River	Hell Gate
ARTHUR KILL	Arthur Kill	Staten Island
PARK AVE.	Harlem River	Park Avenue

NEW YORK DEPARTMENT OF TRANSPORTATION

TYPE OF BRIDGE	LENGTH INCLUDING APPROACHES IN FEET	MAX. SPAN IN FEET	CLEAR HEIGHT ABOVE M.H.W. IN FEET	TRAFFIC FACILITIES Roadway	Footwalks	WIDTH OF CHANNEL OPENINGS IN FEET	DATE OPENED TO TRAFFIC
Bascule	1,398.8	180	27.4	1-53'-0"	2-8'-2"	150	Dec. 3, 1929
Swing	555	227	9	1-19'-6"	2-6'-0"	88.5	Feb. 5, 1903
Fixed	6,021.3	300	125	2-32'-0"	2-8'-0"	250	Aug. 23, 1939
Retractile	345	82	4	1-34'-0"	2-8'-0"	49.5	May 15, 1908
Fixed	341.5	126.5	90	2-32'-0"	None	90	Nov. 15, 1940
Bascule	223	71.5	8.8	1-36'-0"	2-6'-0"	50	Dec. 14, 1910
Bascule	1,400	107	25	2-33'-0"	2-8'-0"	80	April 26, 1939
Fixed	1,806.5	212	25.6	1-42'-8"	2-8'-10"	70	May 14, 1927
Bascule	1,255.5	174.0	35	2-24'-0"	2-2'-0"	140	April 29, 1939
Fixed	145	44.5	7.33	1-72'-0'	2-8'-0"	36.67	Aug. 1, 1931
						31.33	
Fixed	176.2	34.17	4.1	1-62'-6"	2-10'-0"	30.17	Jan. 13, 1931
Bascule	166	123	26.3	1-72'-0"	2-12'-0"	100	Oct. 31, 1925
Bascule	688.9	43	10	1-18'-0"	1-3'-0"	32' Bridge Open	Feb. 26, 1927
Retractile	98	34	4	1-12'-0"	2-1'-11"	30	No Record
Bascule	385	81	11.74	1-53'-0"	2-6'-2"	62	Oct. 29, 1931

TRIBOROUGH BRIDGE & TUNNEL AUTHORITY

TYPE OF BRIDGE	LENGTH INCLUDING APPROACHES IN FEET	MAX. SPAN IN FEET	CLEAR HEIGHT ABOVE M.H.W. IN FEET	TRAFFIC FACILITIES Roadway	Footwalks	WIDTH OF CHANNEL OPENINGS IN FEET	DATE OPENED TO TRAFFIC
Suspension	2-1/3 miles	1,380	143	2-43½			July 11, 1936
Truss Bridge	1,683	383	55	2-40½		n.a.	
Vertical Lift	770	310	55 lowered			n.a.	
			135 raised	2-30½			
Fixed Steel Arch	2,000	800	142.5	3 lanes upper level		n.a.	
				4 lanes lower level		n.a.	Dec. 12, 1936
Vertical Lift	4,022	540	55 lowered	4 lanes			
			150 raised			n.a.	July 3, 1937
Bascule	3,131	131	17.5	6 lanes			June, 1939
Suspension	7,140	2,300	150	6 lanes			April 29, 1939
Suspension	13,410	1,800	142	6 lanes		n.a.	Jan. 11, 1961
Suspension	13,700	4,260	228	6 lanes upper level			
				6 lanes lower level			Nov. 21, 1964

OF THE PORT AUTHORITY

TYPE OF BRIDGE	LENGTH INCLUDING APPROACHES IN FEET	MAX. SPAN IN FEET	CLEAR HEIGHT ABOVE M.H.W. IN FEET	TRAFFIC FACILITIES Roadway	Footwalks	WIDTH OF CHANNEL OPENINGS IN FEET	DATE OPENED TO TRAFFIC
Cantilever	7,109	672	135	4 lane - 42'			June 29, 1928
Cantilever	10,140	750	135	4 lane - 42'			June 29, 1928
Suspension	n.a.	3,500	212	8 lane upper level			Oct. 25, 1931
				6 lane lower level			
Arch	8,460	1,675	150	2-40'			Nov. 15, 1931

BRIDGES

TYPE OF BRIDGE	LENGTH INCLUDING APPROACHES IN FEET	MAX. SPAN IN FEET	CLEAR HEIGHT ABOVE M.H.W. IN FEET	TRAFFIC FACILITIES Roadway	Footwalks	WIDTH OF CHANNEL OPENINGS IN FEET	DATE OPENED TO TRAFFIC
Arch	17,000	1,017		4 tracks			April, 1917
Vertical Lift		558		1 track			1959
Vertical Lift		380	135 raised	4 tracks			1956

GLOSSARY

ABUTMENT—Thick upright support which takes the thrust of an arch.

AIR LOCK—An airtight chamber through which men and material must pass to get to the compressed air chamber of a caisson.

ANCHORAGE—A solid mass, usually of concrete, into which the end of a suspension bridge cable is fastened.

AQUEDUCT—Long conduit for carrying water over land for great distances.

ARCH BRIDGE—Bridge using a curved beam or series of curved stones as its basic load-bearing element.

BASCULE BRIDGE—Derived from the French word for seesaw, the bascule bridge is a contemporary version of a drawbridge over a moat. It has one or two leaves which are weighted heavily on the shore side in order to balance and thereby lift the overwater span.

BENT—Support consisting of two upright beams joined at the top by a crossbar. Bents are sometimes used in trestle bridges and are also used in scaffolding.

BORING—A hole drilled into a riverbed to determine the type of soil on which a bridge pier will rest.

BRACE—A beam or girder placed at an angle to an upright structure for support.

CAISSON—Watertight bottomless box within which men, or more recently automated dredgers, dig through the riverbed to solid rock on which the bridge piers are founded. The caisson is usually left to form part of the bridge foundation.

CANTILEVER BRIDGE—Bridge in which the overwater spans are balanced by heavy weights on the shore sides. There is frequently a suspended span placed between the two counterbalanced overwater arms.

COFFERDAM—A watertight enclosure set in the water and pumped dry to expose the section of a riverbed where a bridge pier is to be built.

COLUMN—Vertical load-bearing support.

DEAD LOAD—The weight of the bridge itself without any traffic moving over it.

DECK—The flat roadway of a bridge.

DRAWBRIDGE—Any bridge that opens for river traffic, whether by lifting, swinging, or being drawn up.

FALSEWORK—Scaffolding to hold up an unfinished span.

GANTRY—A platform of joists used to carry heavy loads or the rails of a traveling crane. Also a gantry crane, which is a crane which moves along rails.

GIRDER—Long beam which can be solid or latticed.

GRADE—Steepness of the slope of the road, usually measured in percent from the horizontal axis.

GUY WIRES—Ropes or cables angling down from the tops of towers to stabilize a structure.

I-BEAMS—A steel joist that looks like the letter I when viewed from the end.

JOIST—A beam, usually supporting the deck or floor or a bridge.

LATTICE—Welded steel bars in a crisscross pattern which connect the strips of metal at the top and bottom of a girder. These lighten the girder.

LIFT BRIDGE—Bridge with a span that goes straight up, working on the same principle as an elevator, with two counterweights and a pulley.

LIVE LOAD—Traffic that crosses a bridge.

MEMBER—Individual component of a structure.

PIER—A bridge support, usually a column with masonry or concrete.

PILING—A large post of timber or steel thrust into a prepared hole or driven down to support weight.

PIN—Solid steel cylinder joining the bridge span to piers.

PNEUMATIC DRILL—Rock breaking tool run by means of compressed air.

SADDLE—A U-shaped mechanism atop a suspension bridge tower, into which the cables are laid.

SPAN—Distance between supports of a bridge.

SPANDREL—The sometimes ornamental space between the curve of an arch and the straight cornice running above it.

STRAND SHOE—Part of an eyebar embedded in the anchorage onto which the cable wires are looped.

STRESS—The force impinging on a structural member, divided by the area that receives the stress.

SUSPENDER—A short vertical cable which enables the forces of the roadway of a suspension bridge to be translated to the horizontal forces of the cables.

SWING SPAN—Bridge in which one or more spans turns on a pivot, usually built on a central pier.

THRUST—Any horizontal force—most often applied to a force exerted by an arch—on a bridge's abutments and into the earth.

TORSION—Twisting force, as in the wringing of a wet cloth.

TRESTLE BRIDGE—Bridge supported by a series of connected pilings or beams.

TRUSS—Rigid framework built up of steel beams which are interconnected and angled like a series of triangles.

TRAVELING WHEEL—A wheel with a pulley which goes back and forth across the length of a suspension bridge to weave the cables. The wheel carries two strands, each wrapped around a pulley and thereby lays four wires on each trip.

VOUSSOIR—A wedge-shaped device forming part of the lower edge of an arch curve.

BIBLIOGRAPHY

Abelow, Samuel. *History of Brooklyn Jewry,* New York, Scheba Publications, 1937.

American City passim.

American Institute of Architects Guide to New York City, New York, 1968.

American Scenic and Historic Preservation Society Reports, 1896-1925.

Architectural Record passim.

Architecture and Building passim.

Armbruster, Eugene L. *Brooklyn's Eastern District,* Brooklyn, 1942.

Atwood, Howland. Materials relating to the first 100 years of the Brooklyn Ferry and its operation.

Beck, Louis. *New York's Chinatown; an historical presentation of its people and places,* New York, 1898.

Berger, Meyer. *Meyer Berger's New York,* New York, Random House, 1960.

Billings, Henry. *Bridges,* New York, Viking, 1956.

Black, Archibald. *The Story of Bridges,* McGraw Hill, New York, 1936.

Brooklyn Bridge Opening Ceremonies Program. New York, 1883.

Brooklyn Eagle, passim.

Brown, Francis Williams. *Big Bridge to Brooklyn,* New York, Aladdin Books, 1956.

Burgess, George and Kennedy, Mike. *Centennial History of the Pennsylvania Rail Road Co. 1846-1946,* Pennsylvania 1949.

Burr, Wiliam. *Report on Design and Construction of Queensboro Bridge 1908.*

Caro, Robert. *The Power Broker,* Knopf, New York, 1974.

Civil Engineering, passim.

Current Literature, passim.

Cook, H. T. *The Borough of the Bronx 1639-1913; its marvelous development and historical surroundings,* New York, 1913.

Davis, Albert E. Address at North Side Board of Trade, 1903.

Engineering Magazine passim.

Engineering News passim.

Engineering News Record passim.

Esquire, December 1964.

Fortune, passim.

Furman, G. *Antiquities on Long Island,* New York, J. W. Bouton, 1875.

Gies, Joseph. *Bridges and Men,* Garden City, Doubleday, 1963.

Haffen, Louis. Borough of the Bronx; A Record of Unparalleled Progress.

Hammond, Rolf. *The Forth Bridge and its Builders,* London, the Shenval Press, 1964.

Harper's Weekly, passim.

Hildenbrand, Wilhelm, Safety of the Brooklyn Bridge; a report, New York, 1902.

Holland, Rupert Sargent. *Big Bridge,* Philadelphia, Macrae Smith Co., 1938.

Hopkins, H. J. *A Span of Bridges; an Illustrated History,* New York, Praeger Publishers, 1970.

Hungerford, Edward. *The Williamsburg Bridge,* Brooklyn Eagle Press, 1903.

Hutton, William R. *The Washington Bridge over the Harlem-River at 181st St. N.Y.C.: A Description of its Construction,* New York, 1889.

Huxtable, Ada Louise. *The Architecture of New York,* Garden City, Doubleday, 1964.

Jenkins, Stephen. *The Story of the Bronx (from Purchase by Dutch to Present)* New York, G. P. Putnam's Sons, 1912.

Jervis, John Bloomfield. *Description of the Croton Aqueduct from the dam at Croton River to the Distributing Reservoir,* New York, G. F. Nesbitt and Co., 1851.

——— *Reminiscences of John B. Jervis, Engineer of the Old Croton,* Syracuse University Press, 1971.

Jewell, J. *Historic Williamsburg; an account of the Settlement and Development of Williamsburg and its environs from Colonial Days to the Present,* Brooklyn, private printing, 1926.

Laidlaw, W. Population of the City of Greater New York, 1920.

Lindenthal, Gustav. *The North River Bridge: How to Finance it, and Who Should Build It,* New York, 1918.

Mayer, G. *Once Upon a City: New York from 1890 to 1910,* New York, Macmillan 1958.

McCullough, David G. *The Great Bridge,* New York, Simon and Schuster, 1972.

Merritt, Raymond H. *Engineering In American Society; 1850-1875,* Lexington, University of Kentucky Press, 1968.

Metropolitan Magazine, New York 1906.

Mock, Elizabeth. *The Architecture of Bridges,* New York, Museum of Modern Art Publication, 1949.

Morris, James. *The Great Port: A Passage Through New York,* New York, Harcourt Brace and World, 1969.

Moses, Robert. *A Half-Century of Achievement, Triborough Bridge and Tunnel Authority Links Five Boroughs, Parks, Beaches and Suubrbs,* New York, 1948.

——— "The Changing City," 1940 (reprinted from *Architectural Forum,* March 1940.)

——— *Public Works: A Dangerous Trade,* New York, McGraw Hill, 1970.

Mumford, Lewis. *Brown Decades,* 1955.

——— *The City in History,* New York, Harcourt Brace and World 1961.

——— *Sticks and Stone,* New York, W. W. Norton 1924.

Municipal Art Society of New York Bulletin on the Manhattan Bridge Plans.

Neville, Anthony E. *Bridges, Canals and Tunnels, The Engineering Conquest of America,* American Heritage Publications in conjunction with the Smithsonian Institution, New York 1968.

Nevins, A. and Krouts, J. *The Greater City: New York 1898-1949,* New York, Columbia University Press, 1948.

New York City Bridge Department. Contract Drawings for the Manhattan Bridge 1903 and 1912.

New York City Department of Public Works. *Modernized Brooklyn Bridge,* May 3, 1954.

New York City Finance Department, *Tunnels or Bridges?*

New York City Department of Bridges. Annual Reports 1905-1974.

New York City Guide. American Guide Series, W.P.A. New York, Random House 1939.

New York Daily Express passim.

New York Herald passim.

New York Times passim.

New Yorker passim.

North River Bridge Co. The North River Bridge at New York, 1895.

Plowden, David. *Bridges; The Spans of North America,* New York, The Viking Press 1974.

Rainey, T. The New York & Long Island Bridge Prospectus, New York, 1884.

Ratigan, William. *Highways over Broad Waters: Life and Times of David B. Steinman, Bridgebuilder,* Grand Rapids, Eerdman, 1959.

Real Estate News, passim.

Riis, Jacob. *The Children of the Poor,* New York, Scribner's, 1923.

Rischer, M. *The Promised City,* Cambridge, Harvard University Press 1962.

Rodgers, Cleveland, *New York Plans for the Future,* New York, Harper and Bros., 1943.

——— *Robert Moses: Builder for Democracy,* New York, N. Holt & Co., 1952.

Roebling, Washington. Report of the Chief Engineer of the New York and Brooklyn Bridges, Brooklyn, New York, 1877.

Roskolenko, Harry. *When I was Last on Cherry Street,* New York, Stein & Day, 1965.

Schuyler, Hamilton. *The Roeblings: A Century of Engineers, Bridgebuilders and Industrialists.* Princeton University Press, 1931.

Schuyler, Montgomery. *American Architecture and Other Writings* edited by Wm. H. Jody and Ralph Coe, Cambridge Belknap Press, 1961.

Science Digest, passim.

Science Newsletter passim.

Scientific American passim.

Smith, Betty. *A Tree Grows in Brooklyn,* New York, Everybody's Vacation Publishing Co., 1943.

Smith, Dorothy Valentine, *Staten Island; Gateway to New York,* Philadelphia, Chilton Book Co., 1970.

Steinman, David. *Bridges and their Builders,* New York, G. P. Putnam's Sons, 1945.

——— *The Builders of the Bridge; the Story of John Roebling and His Sons,* New York, Harcourt Brace & Co., 1945.

——— *Fifty Years of Progress in Bridge Engineering,* American Institute of Steel Construction, 1929.

——— *Miracle Bridge at Mackinac,* Grand Rapids, 1957.

Staten Island Advance passim.

Stokes, Isaac Newton Phelps. *Iconography of Manhattan Island.* 6 Volumes, New York, Robert H. Dodd, 1915-1918.

Sullivan, Patrick G. *The Williamsburg Bridge: The Battle and the Triumph* (master's thesis, St. Joseph College, Brooklyn.)

Syrett, H. C. *The History of Brooklyn,* New York, Columbia University Press, 1944.

Talese, Gay. *The Bridge,* New York, Harper & Row, 1964.

Technical World, June 1914.

Tomkins, Calvin. *The World of Marcel Duchamp 1887* — New York, Time-Life, 1960.

Trachtenberg, Alan. *Brooklyn Bridge, Fact and Symbol,* New York, Oxford University Press, 1965.

Tyrell, Henry Grattan, *Artistic Bridge Design,* Chicago, 1912.

Valentine's Manual 1891.

Waddell, J. A. L. Addresses to Engineering Students, 1911.

——— Aesthetics in Bridge Design, 1917.

Weld, Ralph Foster. *Brooklyn is America,* New York, Columbia University Press, 1950.

Wells, James Lee. *The Bronx and its People; A History, 1609-1927,* New York, The Lewis Historical Publishing Co., 1927.

Whitman, Walt. *Specimen Days in America,* New York, E. P. Dutton, 1906.

Williamsburg Bridge, Souvenir and Official Program for the Celebration of the Elevated Train Service on the Williamsburg Bridge.

World Telegram & Sun passim.

Above—Currier & Ives lithograph c. 1885 showing a panorama of the Brooklyn Bridge and its surroundings.
Back Cover—An 1885 view of the Harlem River bridges.
Both views—COURTESY OF THE NEW YORK HISTORICAL SOCIETY